W9-BRH-752

PAGES
OF
STONE

HALKA CHRONIC

PAGES OF STONE

GEOLOGY OF WESTERN NATIONAL PARKS & MONUMENTS

4: GRAND CANYON AND THE PLATEAU COUNTRY

The Mountaineers • Seattle

THE MOUNTAINEERS: Organized 1906
*". . . to explore, study, and enjoy
the natural beauty of the outdoors."*

© 1988 by The Mountaineers
All rights reserved

No part of this book may be reproduced in any form, or by any
electronic, mechanical, or other means, without permission in
writing from the publisher.

Published by The Mountaineers
1011 S.W. Klickitat Way, Suite 107, Seattle WA 98134
Published simultaneously in Canada by Douglas & McIntyre
1615 Venables Street, Vancouver, British Columbia V5L 2H1
Published simultaneously in Great Britain by Cordee
3a DeMontfort St., Leicester, England LE1 7HD

Cover design by Judy Petry
Layout by Ray Weisgerber
Cover photos: Queen's Garden, Bryce Canyon National Park;
inset—Grand Canyon National Park (Halka Chronic photos)
Frontispiece: Undermined by erosion of soft rock layers near
the floor of Zion Canyon, parts of the thick, massive Navajo
Sandstone break away along vertical joints, leaving the huge
monolith of Zion's Great White Throne. (Ray Strauss photo)

Photos by author unless otherwise credited

Manufactured in Mexico

Library of Congress Cataloging-in-Publication Data
(Revised for vol. 4)

Chronic, Halka.
    Pages of stone.

    Includes bibliographies and indexes.
    Contents: 1. Rocky Mountains and Western
Great Plains — 2. Sierra Nevada, Cascades &
Pacific Coast — [etc] — v. 4. Grand Canyon
and the plateau country.
    1. Geology—West (U.S.) 2. Geology
—Great Plains. 3. National parks and reserves
—West (U.S.) 4. National parks and reserves
—Great Plains. 5. Natural monuments—West
(U.S.) 6. National monuments—Great Plains.
I. Title.

QE79.C47 1984   557.8   82-422
ISBN 0-89886-155-1

2 1
5 4 3 2

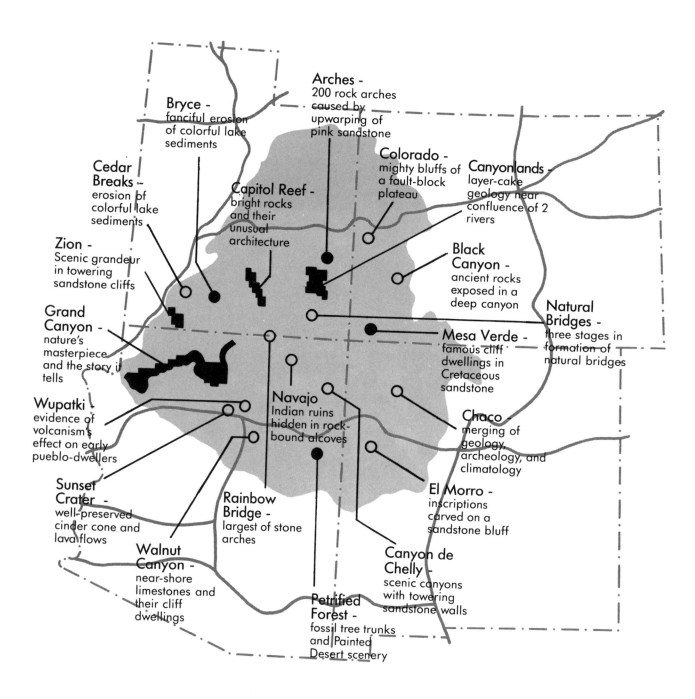

Bryce - fanciful erosion of colorful lake sediments

Arches - 200 rock arches caused by upwarping of pink sandstone

Colorado - mighty bluffs of a fault-block plateau

Canyonlands - layer-cake geology near confluence of 2 rivers

Cedar Breaks - erosion of colorful lake sediments

Capitol Reef - bright rocks and their unusual architecture

Zion - Scenic grandeur in towering sandstone cliffs

Black Canyon - ancient rocks exposed in a deep canyon

Natural Bridges - three stages in formation of natural bridges

Grand Canyon - nature's masterpiece and the story it tells

Mesa Verde - famous cliff dwellings in Cretaceous sandstone

Wupatki - evidence of volcanism's effect on early pueblo-dwellers

Navajo Indian ruins hidden in rock-bound alcoves

Chaco - merging of geology, archeology, and climatology

Sunset Crater - well-preserved cinder cone and lava flows

Rainbow Bridge - largest of stone arches

El Morro - inscriptions carved on a sandstone bluff

Walnut Canyon - near-shore limestones and their cliff dwellings

Canyon de Chelly - scenic canyons with towering sandstone walls

Petrified Forest - fossil tree trunks and Painted Desert scenery

Contents

To Barbara,
and
in fond remembrance of Eddie,
who long ago introduced me to
Plateau geology

Park trails lead to geologic features not apparent from main roads. Here, swirls of ancient sand dunes reappear on eroded surfaces in Zion National Park. Ray Strauss photo.

Preface

My earliest memory of the Plateau country is of a sunny summer's day in the late 1930s. I was driving with my mother from California to Massachusetts—quite an undertaking in those days—on the Lincoln Highway, which ran from San Francisco to the tip of Cape Cod. In southern Utah we diverged from the highway to gaze in delight at the sheer-walled magnificence of Zion and the breathtaking intricacy of Bryce Canyon's sculptured turrets, as lovely a landscape then as it is now.

I'd been in the Plateau country before, though, when my parents accompanied Dr. Byron Cummings—father of southwestern archaeology—into Hopi and Navajo lands east of the Grand Canyon. A procession of cars—for the sake of science but not without the lure of adventure. Not much in the way of roads, I learned later. Turns were taken pulling and pushing each car through mud and sand. Our Studebaker brought up the tail end of the cavalcade. When the expedition bogged down, Mother knelt beside the gearshift and the hand brake to mix my powdered formula and warm it over a Sterno flame. I was six months old.

I slept cozily in a hammock slung between the car doors. My brother slept on the front seat, Mother on the back. Expedition tents sheltered flour and sugar and the precious finds of the archaeologists. My father (and, as he put it, God) slept out in the rain.

The expedition took in Hopi Indian snake dances at mesa-top Walpi. We two little ones stayed with an Indian nurse in Oraibi, our parents blissfully unaware of a measles epidemic there. In Kayenta (in those days pronounced Kay-yen-TAY) we stayed with the Wetherills of trading post fame. Dad and "Old John" Wetherill plunged into a friendship that lasted many decades. From there, my mother rode on horseback to Monument Valley (now a Navajo Tribal Park) with a guide named Harvey Adair—the same Harvey Adair who, when buying a neckerchief at the trading post, requested, "None o' yer gaudy colors. Jes' plain red 'n' yaller." Mother told later of the excruciating discomfort of three days in the saddle, and how she shifted a small pillow to ease the chafing, trying at the same time to absorb and memorize the beauty all around her.

We all went to Grand Canyon a year or so later—I've seen the photographs. We camped on the rim—the *very* rim. No paved roads there, either, and no organized campgrounds.

The road to Monument Valley acquired its pavement during the uranium boom of the 1950–60s—but not before my own children had bounced over its dusty washboards and heard the story of their grandmother's horseback visit. And not before I became reacquainted with the Plateau country, and, indeed, interested in geology, through three summers at the Museum of Northern Arizona in Flagstaff.

And what a place to learn geology! Half the history of the Earth is here—from the dark, hard rocks of Grand Canyon's Inner Gorge, through the canyon's multilayered walls, up the "Grand Staircase" of the Vermilion Cliffs, White Cliffs, and Pink Cliffs in southern Utah, to the volcanic lavas of Sunset Crater, scarcely touched yet by erosion.

Cliff-edged mesas and layer-cake plateaus, each band of rock younger than those below. Rocks alive with color: sandstone and siltstone and limestone. Rocks marked with curly imprints of ancient worm trails, dotted with fossil shells, creased with polygons of ancient mud cracks, festooned with ripple marks, transected by volcanic dikes. Sea-bed sediments high above the sea, dunes of an ancient Sahara, a fossil forest downed by volcanic cataclysm. Rocks worn by rain and snow and frost, deeply gouged by rivers, incised by seldom-flowing streams, etched by wind.

The Grand Canyon was "discovered" by one of Coronado's lieutenants as early as 1541. It was known much earlier by Indians, some of whom lived and farmed in its depths. One of the first Americans to explore this region was Lieutenant J. C. Ives, who was employed by the United States government to investigate the geology of the area. Ives was unimpressed:

> Ours is the first, and doubtless the last, party of whites to visit this profitless locality. It seems intended by nature that the Colorado River, for the major part of its lonely and majestic course, shall be forever unvisited and undisturbed.

But the Grand Canyon was not dead. In 1869 John Wesley Powell, a one-armed veteran of the Civil War, led a daring expedition down the unexplored Green and Colorado Rivers and through the Grand Canyon. Powell firmly believed that a river carrying so much sand and silt and rock rubble would have long ago ground down any waterfalls. Fortunately he was right! The trip took longer than planned, for Powell, though self-educated, was a thorough scientist, and documented every phase of the trip with accurate measurements and vivid descriptions of rocks and river and ruins. Before he and his party emerged from the western end of the Grand Canyon, newspapers across the country brandished obituary notices (about which Powell later joked: "In my supposed death I had attained to a glory which I fear my continued life has not fully vindicated"). Like the Grand Canyon, Powell was very much alive. Partly on the basis of his Plateau country exploration, he went on to urge Congress to establish the United States Geological Survey, an agency whose first task was to further explore the West.

Most of the parks and monuments described in this book by their very nature emphasize geology. A few—Mesa Verde, Chaco, Navajo, El Morro—place emphasis on human history and prehistory, but even in them the rocks stand out boldly, color-fully, often dramatically, as part of the even earlier history of the Colorado Plateau.

To learn about their geology, visit these parklands—both parks and monuments—for all they are worth. See introductory films and slide shows offered by the visitor centers. There you can also find topographic and, for some areas, geologic maps, as well as other literature discussing geology. For finding your way around, the small maps passed out at entrance stations are quite sufficient. With few exceptions, places mentioned in this book appear on these maps.

You'll learn also, as Powell did, by experiencing the geology for yourself, exploring parks and monuments by car and on foot, getting out and seeing with your own eyes, touching with your own fingers. (Note: Collecting rocks, minerals, fossils, and plant and animal material is not permitted in national parks and monuments.) Roads and trails (some with guide leaflets) visit specific geologic features described in this volume. Allow yourself time for exploring these features, as well as for discovering others on your own.

Geology is a logical science, and by pausing long enough to study landforms and rocks, by looking at them carefully, by thinking about what you are seeing, you may be able to interpret some of them for yourself. Use the present as the key to the past. Streams and rivers of the past washed rocks and sand and silt from mountains and highlands, and deposited them on floodplains and deltas, just as modern rivers and streams do. Sand and clay and limy mud deposited in the past have, with time and with the weight of other rocks above them, hardened into sandstone and mudstone and limestone, just as sediments deposited today will in the end, if they are not washed away, become rock. Sands of ancient deserts have counterparts in today's Sahara, or even in dunes here and there in our own Southwest. Winds and flowing waters of the past rippled mud and sand, and raindrops pitted soft surfaces with tiny craters, just as their modern counterparts do. Shellfish of the past, preserved as fossils in rock, have modern descendants. Other animals left other exciting glimpses into their lives—footprints in the sand, now preserved in rock.

Although this book is designed for readers without formal geologic education, I hope it will prove useful to students and professionals as well. In Part 1, geologic terms are printed in **bold type** where first used and, unless their meaning is clear, defined. Most of the terms are defined again in the glossary in the back of the book. In accordance with National Park Service practice, metric measurements are followed by English units in parentheses. Additional reading, suggested at the end of each section, emphasizes articles, books, and maps that can be understood with no more geologic background than that presented in Part 1.

The information presented here is drawn from published geologic literature, from discussions and correspondence with individual geologists who have worked in and near park areas, and from personal observation. I sincerely thank the many colleagues whose work I leaned on, as well as those who helped me more directly. I also express my gratitude to members of and volunteers in the National Park Service, who made my visits to parks and monuments more interesting and exciting, and who in addition reviewed portions of the manuscript. My special thanks go to Richard Hereford of the U.S. Geological Survey, and to Faye and Arthur Frost, who read and critiqued the entire manuscript and offered many helpful suggestions; to Tad Nichols and Ray Strauss, who contributed many of the photographs; and to Luiz Schleiniger, who produced good prints from my often mediocre negatives. My daughters Lucy Chronic and Felicie Williams provided two of the color photographs, and Emily Silver, also my daughter, prepared most of the figures and diagrams.

Writing this book—the fourth of the *Pages of Stone* series—has been a rich and rewarding experience. Not least of the pleasures has been the delightful excuse to hike again the trails of Grand Canyon, to raft the rapids of its great river, to gaze once more on the breathtaking beauty of Bryce, to explore the hidden clefts and chasms of Zion, to prowl the back roads of Canyonlands, and, perhaps most deeply memorable, to stand alone in silent awe, hearing the whispers of the Ancient Ones at the cliff dwellings of Navajo National Monument.

Safety is an important concern in all outdoor activities. No guidebook can alert you to every hazard or anticipate the limitations of every reader. Therefore, the descriptions of roads, trails, and natural features in this book are not representations that a particular place or excursion will be safe for your party. When you visit any of the places described in this book, you assume responsibility for your own safety. Under normal conditions, such excursions require the usual attention to traffic, road and trail conditions, weather, terrain, the capabilities of your party, and other factors. Keeping informed on current conditions and exercising common sense are the keys to a safe, enjoyable outing.

The Mountaineers

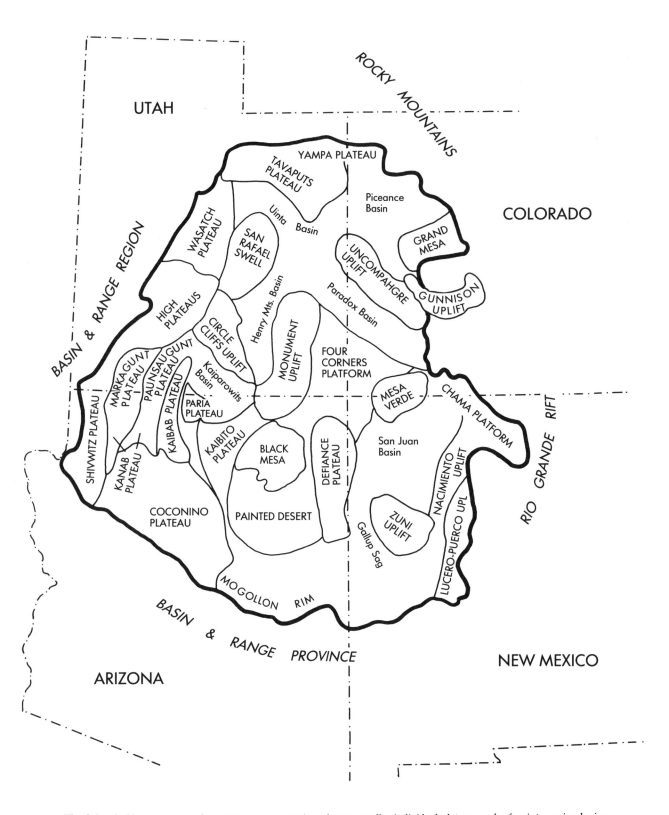

The Colorado Plateau covers a four-state area and consists of many smaller individual plateaus and a few intervening basins.

PART 1

OF EARTH AND TIME

I. The Colorado Plateau

The Plateau country, a land famous for its tranquil beauty, is geologically a raft in a stormy sea. A broad, high tableland, it has steadfastly resisted forces that bent and crumpled the surrounding country and created waves of mountains to the north, east, south, and west. Bypassed by mountain building forces, its rocks, lifted to elevations as great as 3000 meters (10,000 feet), still lie in the horizontal positions in which they formed.

Properly called the Colorado Plateau*, honoring the great river that has carved dramatic canyons through it, this region centers around the Arizona-Utah border but reaches into western Colorado and northwestern New Mexico as well. It is made up of many smaller plateaus—Shivwits, Coconino, Kaibab, Paunsaugunt, Markagunt, and others—each a few hundred meters higher or lower than its neighbors. These smaller plateaus, like the logs of a loosely bound raft, have shifted upward and downward along major breaks in the Earth's crust, so that they now form a sort of "Grand Staircase" from south to north. They vary in character with elevation, with types of rocks exposed at the surface, and with amounts of rainfall or snowfall—usually functions of elevation. Surfaces of these smaller plateaus range from 1500 to 3000 meters (4500 to 10,000 feet) above sea level. Some are separated by river valleys or wide basins, others by lines of cliffs—Pink Cliffs, White Cliffs, Vermilion Cliffs, Straight Cliffs, Circle Cliffs, and so on. All but the highest of the plateaus are arid or semiarid, deprived of water by the great barrier of California's Sierra Nevada, which during much of the year intercepts moisture from the Pacific. In only two seasons does this region receive appreciable moisture: in winter, when storms come in from the northwest, and in summer, when monsoon-type winds sweep in from the south, avoiding the mountain barrier.

Long dry periods between these seasons limit plant growth. In the lowest, hottest parts of the Plateau, sage, saltbush, and other low shrubs dominate, and desert grasses turn green with seasonal rains. Pinyon-juniper woodlands are found at intermediate elevations. And pine, spruce, and aspen forests top the highest plateaus.

Characteristically, deserts develop where evaporation exceeds precipitation. But on the Plateau another factor comes into play: the chemistry of the rocks. Deserts here may be due to gypsum or salt in the rocks, or to clays derived from volcanic ash. Some of these clays swell when they are wet and become soft and crusty when they dry, making plant growth difficult. And without plants to help hold down the soil, unconstrained erosion creates badlands like those of Arizona's Painted Desert.

On other parts of the Plateau, too, especially in canyons and along cliffs that edge different plateau levels, rocks are clearly exposed for long distances. And this means *color*, for the rocks of this region are gaily tinted in shades of red, pink, purple, green, and yellow. It also means a paradise for geologists.

The rocks of the Colorado Plateau may in places be darkened with **desert varnish**, a thin, shiny, blue-black polish of iron and manganese oxides. Desert varnish develops slowly, over many centuries, either from minerals gradually leached from the rock or from soluble material carried in dust and spread thinly, over and over again, by occasional rain. Lichens, too, color rock surfaces with green and orange blotches, or streak them top to

---

* Throughout this book, "the Plateau," spelled with a capital P, refers to the Colorado Plateau as a whole.

In the Painted Desert the very nature of the rocks prevents plant growth and causes true badland erosion. Here, blocks of petrified wood weather out and roll down into the gully. Ray Strauss photo.

Long streaks of black lichens and brown or blue-black desert varnish stain many Plateau Country cliffs. These are in Canyon de Chelly National Monument.

bottom with long black bands that indicate seepage lines where rock remains wetter for longer periods after rain.

Here as elsewhere, erosion works on exposed rock surfaces. In this land of sparse rain and snow, **weathering**, the decomposition and disintegration of rock, proceeds in some cases more rapidly than you might think, providing abundant cargoes of sand and mud for ephemeral streams. As in all desert regions, however, soils are thin and vegetation is for the most part scanty. Landforms on the Plateau therefore *look* different from those found in moist climates. Hardness and softness of different rock layers markedly influence erosion, and one characteristic of the Plateau is the alternation of hard and soft rock layers. Cliffs and ledges of resistant limestone or sandstone alternate with slopes and benches of less resistant shale or mudstone. On both large and small scales, such **differential erosion** adds to the variety of Plateau landscapes.

Outcrop patterns, in a land of horizontal rock layers, follow the contours. Cliffs slowly retreat as they are undermined by erosion of weak layers below. As rocks break away from precipices, individually or in massive rockslides, they lie for many centuries just where they have fallen. Or, tumbling to a canyon bottom, they serve as tools for streams coming to life during sudden thunderstorms, helping the rushing waters gouge steep, angular canyons by hammering rock against rock and rasping rock with sand.

Canyons on the Plateau are steep-walled and nearly barren of vegetation. Without the moisture that lubricates landslides or slow soil-creep of wetter climates, canyon profiles maintain the cliff/slope/cliff alternation described above. At higher elevations, the freezing and thawing of water held in cracks and crevices helps to break down the rock.

In limestone, seeping water trickling through tiny cracks and crevices gradually dissolves part of the rock, slowly enlarging its crevices and shaping corridors and caverns. No very large caverns are known to exist in this region, but twisting passageways are known in several places, as are **sinks** and **solution valleys** caused by the collapse of such passages.

Wind, too, plays a strong role in erosion here. Sweeping across wide plateau surfaces, picking

As shown abundantly in Canyonlands National Park, hard and soft rock layers erode into cliff-slope-cliff (or ledge-slope-ledge) patterns. Cliffs and ledges are commonly sandstone or limestone; slopes are shale and mudstone. Ray Strauss photo.

up sand and small pebbles, it hammers at cliffs and ledges. Particularly in the first few meters above the surface, wind rasps deep clefts and carves strange **honeycomb weathering** patterns, or chisels pinnacles and arches, mushroom rocks and balanced rocks, emphasizing in strong relief the strengths and weaknesses of the rock. Whirling in summer dust devils, wind with its airborne tools quickly frosts rock surfaces (and car windshields) with tiny pits. On broad slopes, wind and rain create **desert pavement,** a thin armor of closely spaced pebbles left behind when finer material—sand and silt—has blown or washed away.

II. Plateau Architecture

Many individual plateaus are outlined by **faults** and **folds. Faults** are breaks in the Earth's crust along which movement has taken place. Most of the faults that separate individual plateaus trend north and south, some in zigzag fashion. Most are steeply inclined **normal** or **reverse faults**; their relative up-and-down motion controls the relative height of adjacent plateaus. Because of cliff retreat in this semiarid country, the actual faults may be some distance out from the base of the cliffs that edge those plateaus.

Faults come in many shapes and sizes. In addition to large faults that separate individual pla-

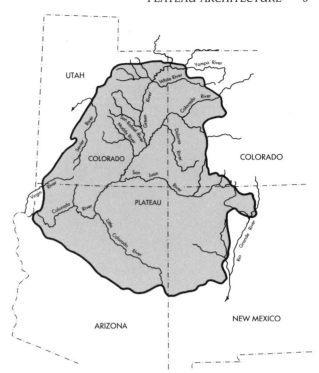

Most of the Colorado Plateau is drained by its namesake the Colorado River and its tributaries, whose winding canyons produce much of the scenic beauty of this area.

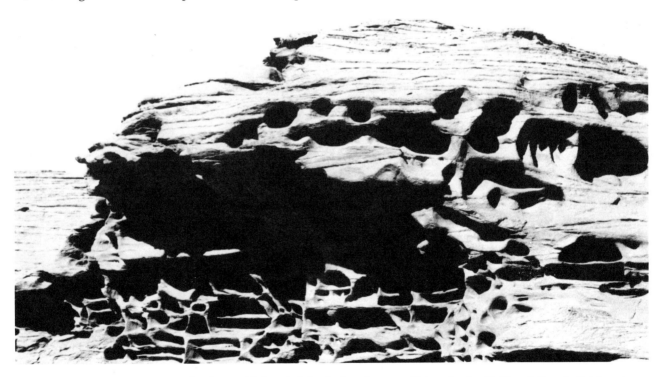

Honeycomb weathering forms as abrasive, sand-laden winds bombard sandstone or siltstone already pitted by solution of salt inherent in the rock.

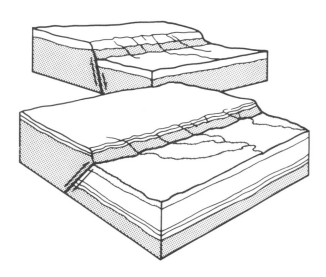

Most of the faults of the Plateau Country are nearly vertical normal and reverse faults.

Where wind blows away smaller rock particles, pebbles concentrate as desert pavement.

teaus, there are thousands of much smaller faults in all parts of the plateau area. The smallest may represent movements of only a millimeter (about .04 inch) or so. Cracks or fractures that show no movement at all—and they are abundant in nearly all rock—are called **joints**.

Folds, on the other hand, are bends rather than breaks. Some plateaus are separated by **monoclines**, the simplest of folds, with one side raised higher than the other. Exposed rocks in the depths of the Grand Canyon show us that some monoclines are surface expressions of faults far below, as shown in the illustration.

Like faults, folds come in many sizes and shapes. On the Plateau we find, in addition to monoclines, gentle downward warps or **synclines**, equally gentle upwarps or **anticlines**, and more or less circular, more or less equidimensional downwarps and upwarps called **basins** and **domes**.

The Plateau, with its well-exposed, straightforward geology, is a geologic textbook—the "pages of stone" that give this series its name. Many geologic features—the fault/fold relationship described above, for instance—can be studied in detail without having to hack away thick underbrush or dig through soil cover. And deep canyons, with their tributaries, give us three-dimensional views. Barren rocks sing out with features that tell us where and how they formed—in seas or on land, on beaches or in lagoons, on deltas or deserts. They even tell us which way rivers flowed or in what direction wind blew in the remote geologic past. What's more, we can

know for sure that in an area so undisturbed, the oldest rock layers are on the bottom, the youngest on top—one of the basic tenets of all geology.

Clearly, geologic architecture gives us our scenery here. The plateaus have risen and fallen, the forces of weathering and erosion—water, wind, and frost—have shaped cliffs and spires, worn little gullies and great canyons, etched and sharpened hard rocks, washed away softer and less resistant ones.

But to understand the basic causes of these changes we need to pause and look at the Earth as a whole, to see what is going on beneath our planet's crust that can influence developments on its surface.

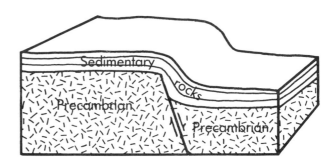

Sedimentary rocks may drape in a monocline over a fault in older rocks below.

In many parts of the Plateau Country, massive sandstone fractures along vertical joints, spilling large angular fragments onto talus slopes below. These are in Capitol Reef National Park. Ray Strauss photo.

Faults frequently consist of clusters of small offsets, as seen here between half-arrows that denote direction of fault movement. Note that evidence of faulting disappears in the dark, soft mudstone at the left. Is this a normal or a reverse fault?

III. The Earth's Interior

Hidden from view, the interior of the Earth long remained a mystery. Very gradually, geologists have found clues to its makeup in earthquake waves and in determinations of its total mass as indicated by its gravity.

From these clues, we have learned that the Earth's **core** is a sphere of almost pure nickel and iron, its center solid but its outer part semiliquid, capable of some degree of plastic movement, as is, for instance, malleable red-hot iron on a blacksmith's anvil. The core is not necessarily smooth-surfaced; recent research shows that it boasts higher mountains and deeper valleys than the Earth's surface.

Outside the core is another thick layer, the **mantle**. It is composed of **basalt**, the same type of black or dark gray rock, rich in iron and magnesium, that occasionally spills out on the Earth's surface as dark gray or black **lava**. As was the case for the core, the inner part of the mantle is solid, the outer part liquid or semiliquid—a seething, slow-boiling, red-hot layer. The mantle is the Earth's stove, heated by the decay of radioactive minerals concentrated there. Convection currents develop in the mantle in the form of huge cells that surge upward in gigantic boils, roll over, and then plunge downward again.

Powered by its own hot stove, the Earth's churning mantle exerts a strong influence on the outermost shell of the Earth, the **crust**. The crust is made up of what we think of as solid rock. On a worldwide scale, there are two major types of crust. The first type, **oceanic crust**, is dark gray or black and heavy with iron and magnesium, very like the mantle in composition. It occurs, as its name suggests, beneath ocean basins. Oceanic crust is quite thin, something on the order of 5 kilometers (a mere 3 miles).

The other type of crust, more familiar to us because we live on it, is **continental crust**. Lighter than oceanic crust in both color and weight, it is composed of minerals low in magnesium and iron. It occurs (as you've no doubt guessed) on continents, extending out to the limits of the continental shelves. Continental crust is relatively thick, about 30 to 40 kilometers (20 to 25 miles).

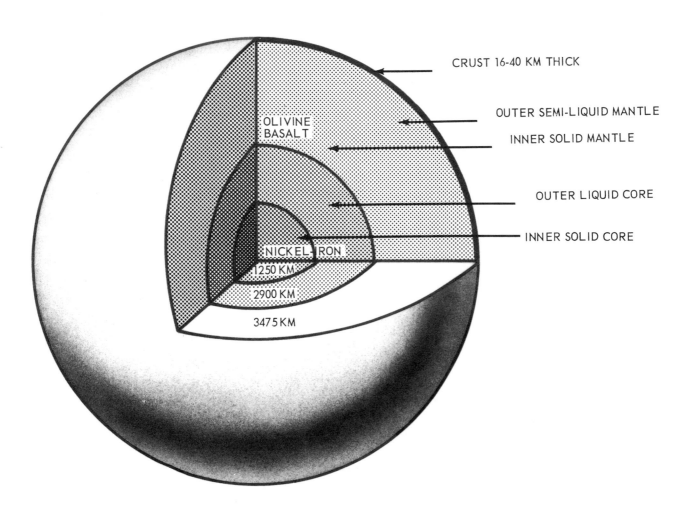

CRUST 16-40 KM THICK

OUTER SEMI-LIQUID MANTLE

INNER SOLID MANTLE

OUTER LIQUID CORE

INNER SOLID CORE

OLIVINE BASALT

NICKEL IRON

1250 KM

2900 KM

3475 KM

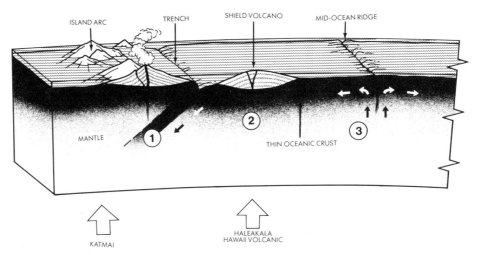

1. Carried along by convection currents in the mantle, oceanic crust is drawn under and melted in subduction zones, forming basalt magma which erupts along island arcs.

2. Shield volcanoes form as plumes of basalt magma rise above isolated "hot spots" below the ocean floor

3. At mid-ocean ridges, basalt magma rises to form new oceanic crust.

IV. Continents on the Move

The idea that continents move, or drift, was born early in this century—a result of recognition of the match in profiles of the continents bordering the southern Atlantic Ocean: Africa and South America *look* like two adjacent pieces of one jigsaw puzzle. And if you juggle other pieces a little bit, particularly if you include their continental shelves, Europe and North America fit together, too. What's more, if you use geologic maps, which show the nature of rocks at the surface, the fit is even better: Adjoining coastlines show many of the same rock types and ages. So some researchers supported the idea that the continents had once been connected, but had broken apart and drifted away from each other.

The catch was that nobody could come up with a plausible theory to explain how parts of the thin, fragile crust could sail off around a globe of solid, or nearly solid, mantle.

In the 1960s, however, geologists developed new views, born, like most scientific views, of the old. The **Theory of Plate Tectonics** answered so many previously perplexing questions that it took the geologic world by storm. Geologists everywhere seized upon it to explain their particular findings in their particular part of the world. In no time the Theory of Plate Tectonics was *in*.

This theory once again sets the continents adrift. But it portrays the crust as coupled with the uppermost part of the mantle to form a stiff layer, the **lithosphere** ("rock sphere"), a layer about 60 kilometers (40 miles) thick under the ocean and about 90 kilometers (60 miles) thick under the continents. Relative to the size of the Earth, that's still pretty thin—hardly more than a film. But like the film of congealed fat on a bowl of chicken broth, it can be rumpled and moved about, or even broken, by stirrings in the mantle.

The lithosphere, and therefore the crust, is made up of a dozen large **plates** fitted together like the plates of a turtle's shell, with a bevy of smaller plates squeezed in here and there among the large ones. Between the plates are suture lines that in some cases are the submerged volcanic mountain chains of **mid-ocean ridges**, and in other cases are deep-sea **trenches** or magnificent high-rise ranges like the Alps and the Himalaya. All these suture-line areas are beset with frequent earthquakes and periodic volcanic activity, to the extent that maps showing earthquake loci or volcanoes distinctly outline the tectonic plates.

The theory also states that new crust is constantly coming into being along the mid-ocean ridges that thread their way down the Atlantic, through the Indian Ocean, and across the Pacific. The ridges bear along their summits narrow volcanic fissures where molten rock or **magma** boils up from the mantle and hardens into new oceanic crust. Created in two narrow bands, the new crust slides apart as still newer lava erupts along the

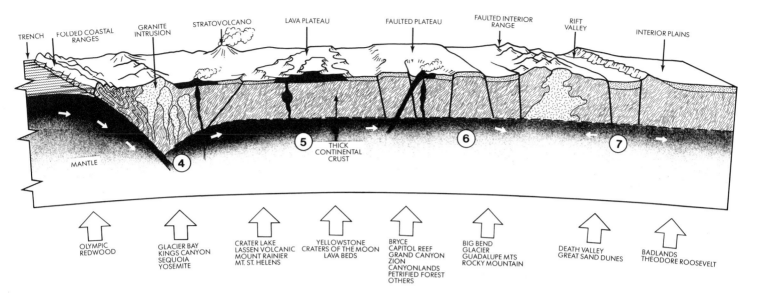

TRENCH   FOLDED COASTAL   GRANITE   STRATOVOLCANO   LAVA PLATEAU   FAULTED PLATEAU   FAULTED INTERIOR   RIFT   INTERIOR PLAINS
RANGES   INTRUSION   RANGE   VALLEY

MANTLE   THICK CONTINENTAL CRUST

OLYMPIC   GLACIER BAY   CRATER LAKE   YELLOWSTONE   BRYCE   BIG BEND   DEATH VALLEY   BADLANDS
REDWOOD   KINGS CANYON   LASSEN VOLCANIC   CRATERS OF THE MOON   CAPITOL REEF   GLACIER   GREAT SAND DUNES   THEODORE ROOSEVELT
SEQUOIA   MOUNT RAINIER   LAVA BEDS   GRAND CANYON   GUADALUPE MTS
YOSEMITE   MT. ST. HELENS   ZION   ROCKY MOUNTAIN
CANYONLANDS
PETRIFIED FOREST
OTHERS

4. Melting of continental crust along a subduction zone creates granitic magma, which may cool slowly, below the surface, in batholiths that are later bared by erosion. or the magma may erupt explosively to form stratovolcanoes.

5. Flood basalts, exceptionally fluid in nature, rise above "hot spots" below continental crust.

6. Interior ranges push upward in response to compression along the distant continental margin.

7. Rift valleys are tensional features where the crust drops as neighboring areas are pulled apart.

fissures. This continuous replenishment makes of the ocean's floor two broad conveyor belts moving slowly and in opposite directions away from each mid-ocean ridge. Movement of the conveyor belts comes about because plates on each side of mid-ocean ridges are dragged apart by slow-moving but powerful convection currents stirring the semimolten part of the mantle. The whole process, which we now call **sea-floor spreading**, is a vital part of the Theory of Plate Tectonics.

To make up for new crust born along mid-ocean ridges, old crust is destroyed near the continents. Where a continental plate collides with an oceanic plate, the oceanic plate, made of heavier material, goes under. It is overridden by the lighter (though thicker) continental plate, and is then drawn down, **subducted**, to depths where it remelts and once more becomes part of the mantle.* Where

both colliding plates are continental, mountains are pushed up, foreshortening the crust. This has happened, for instance, where India, once a small continent in itself, bumped into Asia with enough force to push and crumple and thrust up the Himalaya.

In the oceanic plate–continental plate type of collision, volcanism is likely to occur near the line where **subduction** takes place, with magma rising from two sources: remelted oceanic crust, which gives rise to dark, iron-rich magma; and portions of the leading edge of the continental plate, drawn down and remelted along with the oceanic crust. The more buoyant continental material, once melted, may push upward through the crust near the edge of the continent. If such magma rises only partway to the surface, and cools while it is still at considerable depth, it hardens—crystallizes—slowly into great masses of **granite**. Where it breaks through to the surface, volcanoes form, with light-colored lava unlike the dark lava derived from oceanic crust or directly from the mantle.

---

* Rock melts at temperatures of 1100 to 1200 degrees Celsius, or about 2000 to 2200 degrees Fahrenheit.

NORTH   AMERICAN   PLATE

MID-ATLANTIC RIDGE

PACIFIC PLATE

EAST PACIFIC RISE

SOUTH AMERICAN PLATE

NAZCA PLATE

With formation of new crust along the Mid-Atlantic Ridge, the North American Plate has moved westward (arrow) across the East Pacific Plate, now mostly hidden under western North America. Mid-ocean ridges are offset by numerous transform faults.

The North American Plate, like most other large plates, is partly oceanic (its eastern half, under the Atlantic Ocean) and partly continental (its western half). For the last 60 or 70 million years, as new crust formed along the Mid-Atlantic Ridge and the Atlantic's sea floor spread, the North American continent drifted westward and southwestward at a rate of about 4 centimeters (less than 2 inches) per year. Long ago its westward drift put it on a collision course with the East Pacific Plate and some smaller plates clustered along it.

Because the western part of the North American Plate is composed of continental crust, lighter in weight than the oceanic crust of the East Pacific Plate, the North American Plate overrode the East Pacific Plate, forcing it downward to remelt in the mantle. Eventually the North American Plate overrode almost all of the East Pacific Plate, even covering large portions of the East Pacific Rise, the Pacific's mid-ocean ridge. As this happened, the advancing edge of the continent collected a number of large islands or microcontinents, adding them to its western margin, so the continent gradually became wider in a westward direction.

But just where does the Colorado Plateau fit into this picture? A perplexing question—one for which we don't yet have a satisfactory answer. The Plateau seems to have resisted the forces involved in mountain building, and to have maintained its integrity right through the westward drift and the collision of the North American and East Pacific Plates, as well as through several multidirectional driftings that occurred earlier in the history of the crust. On the Plateau, rock layers ranging in age from 600 million to a mere 6 million years are still in the horizontal position in which they were deposited. Geophysical studies, which measure the thickness of the crust by the velocity of earthquake waves that pass through it, show that the crust is thicker under the Plateau than under surrounding areas, a factor that probably contributes to its raftlike integrity and to its longevity.

The uplift that raised the Plateau, with all its segments, to its present altitude seems to have resulted from general uplift of the western part of the continent, perhaps due in turn to the upwelling of a heat cell in the mantle, related to the way the continent has drifted westward over the East Pacific Rise. The Plateau's normal faults and simple monoclines are due to tension, the pulling apart that might be expected when part of the crust is domed upward. A few reverse faults, though, show compression. Some data suggest that the Plateau is rotating horizontally, like a wheel on its side, which could bring about both compression and tension.

By and large, the Plateau does seem to have been subjected to the same strains and stresses that governed faulting and collapse in the Basin and Range region to the west, with its faults more or less paralleling most of the adjacent ranges. However, the thick, strong crust of the Colorado Plateau obviously responded differently to these stresses than did the Basin and Range region.

V. Rocks and Minerals

Geologists recognize three main classes of rocks, all of them represented on the Colorado Plateau. Since sedimentary rocks are by far the most common type in this area, we'll look at them first.

• **Sedimentary rocks** form from the broken fragments or dissolved minerals of other rocks, transported and deposited by water, wind, or ice. Nearly always, sedimentary rocks are layered or **stratified**, which makes them easy to recognize. (Lava flows and falls of volcanic ash may be stratified, too, though they are classed as volcanic rocks.) Nearly always, the stratification was originally horizontal or nearly so.

Sedimentary rocks range from **conglomerate** made of bouldery gravel to very fine-grained **claystone**. They also include limestone and dolomite made not of rock fragments but of **calcium carbonate** and **magnesium carbonate** precipitated chemically or derived from literally billions of plant and animal shells. Sedimentary rocks frequently contain **fossils**, remains or traces of animals and plants that lived and died when the rock was forming. Sandstone and siltstone deposited by flowing water or wind may be **cross-bedded**, with fine laminae slanting at an angle to the stratification. Or they may preserve **ripple marks** and **mud cracks** that formed on their surfaces as they were deposited, and that are important clues to their mode of origin.

Sedimentary rocks are classified by grain size and composition. They are also classed as **marine** if they are deposited in the sea, and **continental** if

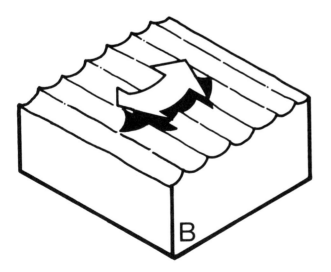

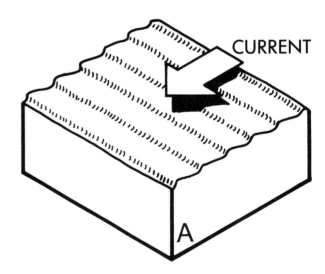

Ripple marks are formed by moving water or wind. Current-formed ripples (left) differ in shape from those formed by oscillating water.

Scalloped by wind, these ripple marks grace a sandstone slab in Capitol Reef National Park.

Long, sweeping cross-bedding tells of windblown sand that accumulated as sand dunes. Crumpled layering at the top of this sandstone shows where sand avalanched down a steep dune face.

they are deposited by rivers, lakes, glaciers, or wind.

• **Igneous rocks** originate from molten rock material known as magma, which rises from as much as 300 kilometers (200 miles) below the surface. Igneous rocks are further divided into two groups: those that cool and harden very slowly deep below the surface, called **intrusive igneous rocks**, and those that harden more rapidly at the surface, called **extrusive igneous rocks** or just **volcanic rocks**.

Long, slow cooling promotes crystal growth, so intrusive igneous rocks are generally grainy, with easily distinguished, tightly packed mineral crystals. Though they vary in composition, these grainy rocks can all be called **granite**. Granite is relatively rare in the Plateau country, but can be found in the depths of the Grand Canyon, in Black Canyon of the Gunnison National Monument, and in Colorado National Monument. It also occurs in the Plateau region in the centers of **laccoliths**, hills or mountains of sedimentary rock pushed up into domes by rising magma forcing a way between sedimentary rock layers.

Thin sheets of intrusive igneous rock may also be sandwiched between layers of sedimentary rock as **sills**, which parallel the stratification of the sedimentary rock, and as **dikes**, which cut across rock layers. Both sills and dikes are crack fillers, either injected with enough pressure to force the older rocks apart, or flowing into open joints or faults. They may branch from conduits of volcanoes, and may have the same chemical and mineral composition as lava erupted at the surface. Some are **feeder dikes** of volcanoes, and some are simply hardened, crystallized fluids that collected in shrinkage cracks of other intrusive igneous rock. Since sills and dikes, as well as small laccoliths, cool more rapidly than the larger

masses associated with major mountain ranges, their crystal grains are smaller.

Volcanic rocks are fairly common on the Colorado Plateau. Basalt, appearing in lava flows and **cinder cones**, is dark in color and quite fluid or foamy when it erupts. Some basalt lava flows, having followed stream valleys, are long and narrow. Others spread out over sizable parts of the Plateau.

Lighter in color than basalt, **silicic volcanic rocks** are much less fluid when they erupt. They form short, thick lava flows, mound into **lava domes**, or plug up a volcano's **vent** until they are exploded away. Alternating with layers of volcanic ash resulting from explosive eruptions, silicic lava flows may pile up into sizable **composite volcanoes** or **stratovolcanoes**, of which there are several on the Colorado Plateau.

• **Metamorphic rocks** form from preexisting sedimentary or igneous rocks that are acted upon by intense pressure and heat, usually stemming from deep burial or from folding and mountain building. Their grains may simply fuse together, in which case their origins can still be recognized. Such rocks may almost melt into new magma, and recrystallize completely, so that it's quite difficult to deduce their original nature. In some, for instance, swirly dark and light banding suggests that they once were stratified, but whether the layers were sandstone and siltstone, or lava and ash flows, cannot be determined.

The two main types of metamorphic rock are **gneiss**, a coarse-grained, usually banded rock very like granite, and **schist**, a finer-grained rock that splits along irregular parallel planes because of its large component of the platy mineral **mica**.

Very old metamorphic rocks—both gneiss and schist—are found in Grand Canyon National Park, Colorado National Monument, and Black

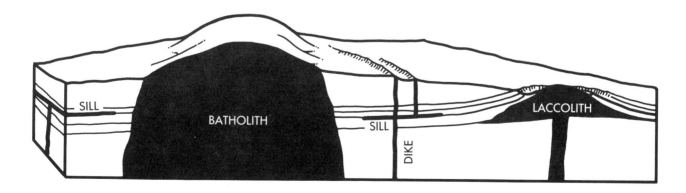

Molten rock—magma—that cools below the surface becomes intrusive rock. Intrusions come in many shapes and sizes, and may later be bared by erosion, as shown here.

Canyon of the Gunnison National Monument. They almost certainly underlie, along with granite, the rest of the Colorado Plateau.

* * * * *

All rocks are made of **minerals**, naturally occurring substances that have definite chemical makeups and often definite and characteristic ways of crystallizing or of breaking. By identifying one or two common minerals in each kind of rock, geologists refine rock descriptions, as for instance when they speak of **biotite schist** or **quartz sandstone**.

Even without a course in geology, you probably can already recognize quite a number of minerals:

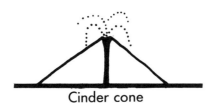

Cinder cone

Spatter cone

Shield volcano

Stratovolcano

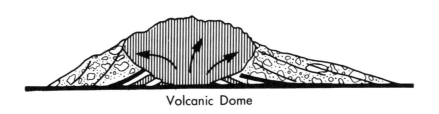

Volcanic Dome

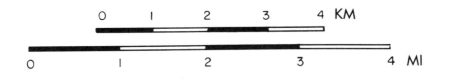

Volcanoes of several types occur on the Colorado Plateau.

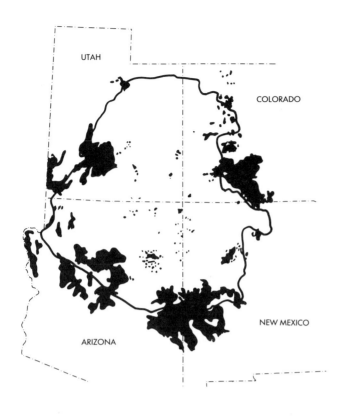

Volcanic rocks—lava flows, cinder cones, and stratovolcanoes— cover large areas near the Plateau margins. Smaller dots are eroded centers of older volcanoes.

quartz, mica, native gold, gemstone minerals like ruby and diamond, semiprecious minerals like turquoise and agate. **Gypsum** and ordinary table salt are familiar minerals, too. **Quartz** and the light pink or gray members of the **feldspar** group are by far the most common of the rock-forming minerals. **Calcite** is the chief component of limestone and marble. Not many collector-quality minerals occur on the Plateau, and collecting of any kind is prohibited in national parks and monuments anyway, so minerals won't be emphasized in this book.

The almost gaudy rocks of the Plateau country owe their color variations to their mineral content. Tiny amounts of iron oxide—the minerals **hematite** (reddish brown) and **limonite** (mustard yellow), the same substances as red and yellow rust—give the rocks varying hues of pink and pale buff. Larger amounts of these same minerals produce deeper reds and brighter yellows. Many of these hues come about when groundwater percolates through rock that contains fine particles of black mica, hornblende, or other nonoxide iron minerals. When unoxidized, as when rocks form under anoxic (no oxygen) conditions, iron imparts a gray-green or purplish tone.

VI. Geologic Dating

Geologic time goes back to the creation of the Earth about 4.6 billion years ago—a pretty sizable chunk of time. But the rocks of the Earth's crust can only be studied back to about 3.8 billion years ago, when the oldest rocks now known were formed. The oldest rocks of the Colorado Plateau are the metamorphic and igneous rocks exposed in the depths of the Grand Canyon and the Black Canyon of the Gunnison, and in Colorado National Monument. They are more than 2 billion years old. Since they formed, continents have more than once drifted together and broken apart. Mountains have formed and worn away and formed again, only to wear away once more. Life has burst forth, first in the seas and then on land, slowly evolving to the plants and animals we know today.

But how do we know this? How can we recognize such immensities of time? How can we "date" rocks? Obviously, days, weeks, and years don't mean much against the immensity of geologic time, in which we deal with millions and billions of years. But we *would* like to know how many millions, or in some cases, thousands, of years ago individual rock units formed.

Early dating methods relied on two things: the position of rocks relative to each other (we've already seen that in undisturbed sequences the oldest rocks are at the bottom, the youngest on top), and fossils, relics or traces of life preserved in rocks. Dating by relative position or by fossils doesn't tell us exactly how long ago a given rock unit formed, but it does tell us which rock is older or which is younger—as long as the rocks are in approximately their original orientation. These techniques, however, work only for sedimentary and volcanic rocks, which always form from bottom to top. Intrusive igneous rocks tend to push up from below.

The use of fossils to date rock units came at a time when the evolution of plant and animal life had not yet been demonstrated. With Charles Darwin and a new understanding of evolution, earlier findings that certain recognizable fossils could be found in certain recognizable rock layers took on a new, two-way meaning: Successive fossil-bearing layers illustrate evolution, and evolution can be used to define the ordered sequence of rock layers on a worldwide basis.

With these dating techniques, haphazard though they were at first, geologists put together a worldwide geologic calendar showing the relative ages of rocks, with names for months, weeks, and days of geologic time. In this calendar, the

GEOLOGIC TIME

ERA	PERIOD	EPOCH	AGE IN YEARS
CENOZOIC Age of Mammals	QUATERNARY Q	HOLOCENE Q	
			— 10,000 —
		PLEISTOCENE Q	
			— 2 million —
	TERTIARY T	PLIOCENE Tp	
			— 5 million —
		MIOCENE Tm	
			— 24 million —
		OLIGOCENE To	
			— 37–38 million —
		EOCENE Te	
			— 55–57 million —
		PALEOCENE Tp	
			— 63–66 million —
MESOZOIC Age of Reptiles	CRETACEOUS	K	
			— 138–144 million —
	JURASSIC	J	
			— 205–208 million —
	TRIASSIC	℞	
			— 240–245 million —
PALEOZOIC Age of Fishes	PERMIAN	Pm	
			— 286–290 million —
	PENNSYLVANIAN	P or ℞P	
			— 320–330 million —
	MISSISSIPPIAN	M	
			— 360–365 million —
	DEVONIAN	D	
			— 408–410 million —
	SILURIAN	S	
			— 435–438 million —
	ORDOVICIAN	O	
			— 500–505 million —
	CAMBRIAN	Є	
			— 570 million —
PRECAMBRIAN P-Є	ORIGIN OF LIFE		2.5 billion
	ORIGIN OF EARTH		4.6 billion

largest units, geologic months, are called **eras**. Eras are divided into **periods**. For detailed work, periods are divided into **epochs**. Names for major time units are shown in the chart on the opposite page. Epoch names, which vary from continent to continent, are given here for American Tertiary and Quaternary Periods only.

Geologists have now learned how to determine the absolute age of rocks with some degree of accuracy. This they do by measuring the decay of radioactive minerals such as uranium, carbon-14, or radioactive potassium-40, and their abundance relative to their own decay products. This technique is known as **radiometric dating**, and it gives us reasonably accurate dates of origin, in years, for tested rocks and minerals. By combining radiometric dating with the older calendar of geologic time, we now have a more accurate calendar on which to base our knowledge of the history of the Earth.

Another newly developed technique relates natural rock magnetism to the pieced-together history of reversals in the Earth's magnetism, when the north and south magnetic poles switched their positive and negative polarity. This technique is called **paleomagnetic dating**. The pattern that develops from switches in polarity (which are not at all regular with time), determined by studies of sea-floor basalts in which iron minerals, oriented by the Earth's magnetism, were "frozen" into the rock, can also be added to the geologic calendar. Polarity reversals are now known back to about 170 million years.

One more type of dating, applicable to quite young, quite recent deposits, relies on variations in the widths of tree rings. Used a great deal by archaeologists to date prehistoric ruins and artifacts, it is equally useful for dating prehistoric volcanic eruptions, floods, and other events that preserved trees or fragments of wood.

Except for very recent events, instead of using B.C. or A.D., geologists denote time as B.P.—"before present"—or as so many years "ago." Their figures are by their very nature approximate. A geologic event that happened 100 million years ago will not be significantly older next year.

Because of the way the geologic calendar developed, boundaries between its time divisions reflect notable changes in the history of the Earth and its inhabitants. The era divisions were meant to indicate stages in the development of life, from "early life" (Paleozoic) to "middle life" (Mesozoic) to "recent life" (Cenozoic). These three eras are also popularly called the Age of Fishes, the Age of Reptiles, and the Age of Mammals, named for their dominant animal types. When the Precam-

brian Era was first defined, life was thought not to have existed before Paleozoic time. We know now that both plants and animals *did* exist then, in highly complicated forms. The ancestors of all our major groups, as well as some groups without modern descendants, had already evolved, but did not secrete the hard shells and skeletons likely to be preserved as fossils.

Although the rock record is more or less continuous, with gaps in the record in one area filled in somewhere else, there do seem to be sudden breaks in the orderly progression of plant and animal life. Some groups seem to go on forever, but others gradually or abruptly became extinct.

Extinctions, particularly the demise of the dinosaurs at the end of the Mesozoic Era, have puzzled geologists for decades. A possible answer to the puzzle is now receiving quite a bit of publicity: The element **iridium**, which is rare on the Earth's surface but abundant in meteorites and in the Earth's mantle, has been found in many places right at the Mesozoic-Cenozoic (Cretaceous-Tertiary) boundary. This concentration suggests that the sudden obliteration of dinosaurs and other animal groups that vanished from the Earth at the same time may have been brought about by the impact and explosion of extraterrestrial bodies—possibly large, iridium-rich asteroids. As dust thrown up by the impacts and explosions spread out in the upper atmosphere, the world would become shadowed and cold. Plants would wither and die, interrupting food chains of many animals. Recent research even suggests that the animals that died were those dependent on food chains involving living plant matter, while those that survived the great extinction were dependent on food chains not involving living plants.

Other geologists point out that iridium layers, which have now been found at other earlier-recognized boundaries corresponding with other extinctions, may have been brought about by unusually intense bouts of volcanism, which may have heated up the atmosphere and thrown quantities of dust and perhaps toxic acids into the stratosphere. Extinctions great and small seem to occur in more than the usual abundance about every 26 million years, however—a periodicity that supports an extraterrestrial rather than an Earth-born cause.

It's worth mentioning that breaks in the rock record, including those coinciding with time divisions of the geologic calendar, are called **unconformities**. They commonly take the form of erosion surfaces that bevel or channel older rock layers, with younger rock layers then deposited

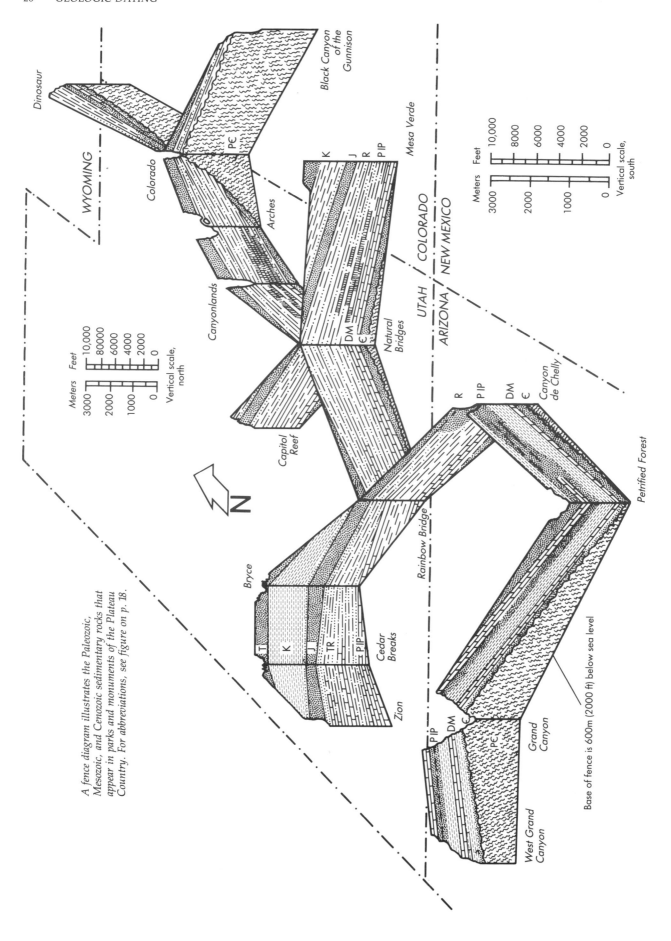

A fence diagram illustrates the Paleozoic, Mesozoic, and Cenozoic sedimentary rocks that appear in parks and monuments of the Plateau Country. For abbreviations, see figure on p. 18.

on top of them. Unconformities and the breaks in the fossil record that they entail are due to times of widespread uplift and erosion, or to environmental changes, or, going along with the theory explained above, to asteroid collisions. Unconformities occur on small scales, too, representing short or localized intervals when sedimentation ceased and erosion held sway. Gaps in the rock record are commonplace; there is no place on Earth where we have a complete, uninterrupted record of all geologic history.

VII. Maps and Figures

Photographs have become an integral part of geologic literature. However, they do not completely edge out maps, diagrams, and sketches, which sometimes do a better job of clarifying basic geologic relationships. Let's look at the types of geologic illustrations used in this book.

Geologic maps show rock units—specific, named, easily recognizable units called **formations** or **groups** of formations—that occur at the Earth's surface or just under the loose soil and rock debris on the surface. Geologic maps are the prime product of most geologic field research. Plotting rocks visible in **outcrops** or revealed by shallow digging, geologists of the United States Geological Survey (the USGS) have worked many patient weeks and months to give us detailed geologic maps of most of the United States. Survey maps, in color, in 7½- and 15-minute quadrangles that correspond to topographic quadrangle maps, can be obtained from USGS map offices in Reston, Virginia; Denver, Colorado; and Menlo Park, California. Geologic maps of larger areas—whole states or some individual national parks and monuments—are available there also. Park and monument geologic maps, if they exist, can also usually be purchased at visitor centers.

These maps show to the practiced eye not only the formations and groups, but the faults, folds, and **dip** (downward slant) of, in particular, sedimentary rock layers. A much-used geologic term, dip is measured in degrees below horizontal, in the direction of maximum slope of a rock layer. On maps it is represented by the dip symbol ⟋ **28**, the small line pointing in the direction of dip. The number of degrees may or may not be shown.

Cross sections diagrammatically slice open the rocks below the surface to give a picture of what geologists think is there. **Block diagrams** combine cross sections with perspective views of surface features, showing what a block of the Earth's crust would look like if it could be lifted out of its natural surroundings. Cross sections and block diagrams are usually easier to understand than maps, and are good ways to show geologic features at a glance. They are made from maps or, where rock layers are well exposed, from the rocks themselves. Block diagrams are particularly useful for interpreting geologic and topographic changes resulting from faulting, folding, uplift, and erosion. In both cross sections and block diagrams, the vertical dimension may be exaggerated in order to show the succession of rock layers more clearly. By convention, unconformities are shown by wiggly lines: ∿.

On page 20 is another type of geologic illustration derived from cross sections: a **fence diagram**. Made from many cross sections put together in their correct geographic position, fence diagrams assemble a *lot* of data in one picture. This one shows the position, age, and general composition of rocks in national parks and monuments discussed in this book, and is a useful guide to the geologic history of the entire Plateau.

Another type of illustration particularly suited to the Colorado Plateau is the **stratigraphic diagram**—a picture summarizing the rock layers exposed in many outcrops, as they would appear if piled on top of one another in the order in which they were originally deposited (oldest at the bottom, youngest on top). Stratigraphic diagrams can also show how rock layers weather and erode—some as slopes, some as ledges and cliffs—a feature that makes them extra useful on the Colorado Plateau, where a lot of scenery results from these very characteristics. Under natural conditions, individual rock layers may not everywhere erode in exactly the same way, so one must make some allowances in trying to match up scenery and stratigraphic diagrams.

VIII. The Plateau Story

With all these basic concepts in mind, we are ready for a quick overview of the geologic history of the Colorado Plateau. Let's take it era by era.

Precambrian Era. The history of the Plateau begins in Precambrian time, more than 2 billion years ago, when ancient gneiss and schist now visible in the depths of the Grand Canyon and in the Black Canyon of the Gunnison were formed. These rocks came into being as yet older rocks—perhaps sandstone and shale, perhaps lava and volcanic ash—were heated, crumpled, and recrys-

Tilted into vertical position, Grand Canyon's Vishnu Schist, above the hikers, is beveled at the top and covered with horizontal Cambrian strata.

Dark Precambrian gneiss forms the formidable walls of the Black Canyon of the Gunnison.

tallized. These predecessors appear to have been immensely thick, on the order of tens of kilometers. Squeezed and tilted into nearly vertical position by mountain-building forces, they underlie most of the continent.

Later, large masses of granite magma pushed up into them. Then, the whole landscape was gradually beveled by erosion to a nearly horizontal surface.

In parts of the Plateau, including Grand Canyon, this horizontal surface was overlain by younger Precambrian rocks that even now are clearly recognizable as sedimentary and volcanic in origin. There are layers of conglomerate, sandstone, and shale, as well as dark lava flows. Eventually these rocks were broken and tilted by block faulting, and then beveled once more during a long

erosional episode at the end of Precambrian time.

Paleozoic Era. The Paleozoic Era was one of calmness and relative sameness. The beveled land rose and fell, more than once, just enough to allow a shallow western sea to advance and retreat across the nearly horizontal landscape. At times sand dunes swept what is now the Colorado Plateau. But in this area, sedimentary deposits on the whole were thin—a few hundred meters during Cambrian time, little or none (none are preserved, anyway) during Ordovician and Silurian time, next to none during Devonian time, more during Mississippian, Pennsylvanian, and Permian time. A couple of thousand meters at most—against the many thousands of meters in a deep marine trough to the west.

Horizontal cliffs, ledges, and slopes stripe Grand Canyon's walls. Paleozoic rocks in the canyon range from Cambrian to Permian.

Fossils of many kinds are found in Plateau Country rocks. These snails and clams are from the Permian Kaibab Limestone.

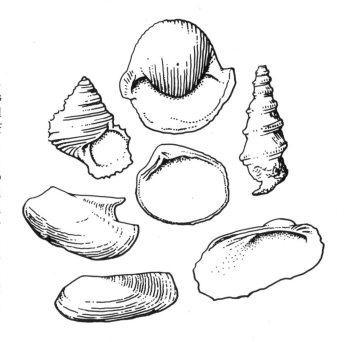

The shallow Paleozoic seas were lively places. Sea-dwelling animals that had their beginnings well back in Precambrian time drifted and crawled and swam and burrowed in the soft sediment of the sea floor: brachiopods, trilobites, corals, clams and snails, feathery bryozoans, and fishes, the first vertebrates. What is more, they began to grow their own protection, hard shells that would be preserved as fossils. Beginning in Cambrian time, plant and animal shells added significantly to the sediments, building layers of limestone. Preserved as fossils, the shells now give us glimpses of evolving life. By Devonian time, plants and some creatures of the sea had crept onto the land. In Permian time their descendants left footprints in damp dune sands.

In summary, the Paleozoic Era was a peaceful time: no volcanism, no mountain building, only the slow accumulation of sediments across an almost horizontal surface.

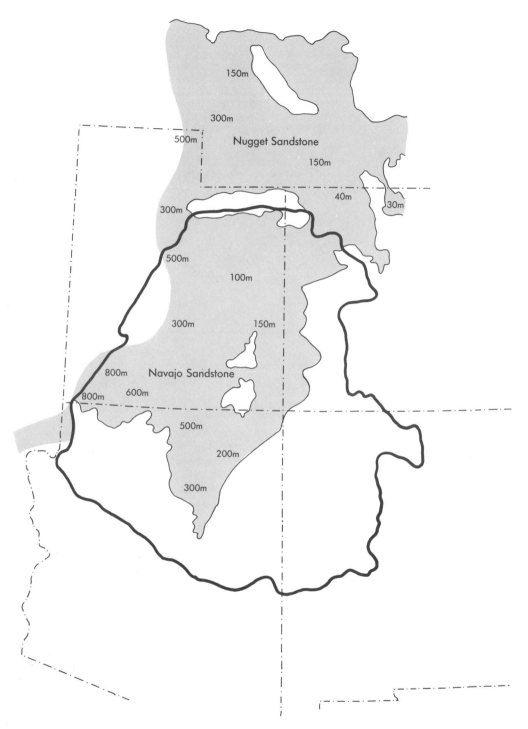

The Navajo Sandstone and its northern equivalent, the Nugget Sandstone, were deposited on a Triassic-Jurassic desert comparable to today's Sahara. Both formations thicken westward to more than 500 meters (1500 feet).

Mesozoic Era. With scarcely an interruption, the land rose at the end of Paleozoic time, and the western sea drained away. Mountains that had formed in Colorado and northern New Mexico in Pennsylvanian time shed sediments onto the flat Plateau region, building floodplains and deltas.

Now and then, nearby volcanoes belched volcanic ash. Well to the west, molten magma pushed up to form the granites of today's Sierra Nevada, a range that then as now cut off moisture swept eastward from the Pacific. A great Sahara developed across the Plateau region, its high-piled

Badlands of Triassic and Jurassic mudstones characterize many parts of the Plateau.

dunes giving us many of the scenery-making rocks of the region. Just as today's Saharan dunes creep across the Nile floodplain, the dunes of this ancient Sahara advanced across older floodplains, covering and in places distorting, with their weight, the soft, slippery river muds.

Across both desert and delta wandered the newly dominant reptiles, among them the dinosaurs, for nearly 200 million years the monarchs of the land. In forests and swamps, trees lived and died, some perhaps perishing in showers of volcanic ash and great mudflows resulting from volcanic eruptions. And as a shallow sea again swept this area, large marine reptiles and elegantly spiraled ammonites, shelled relatives of squids and octopuses, swam above the fine gray mud of the sea floor.

Toward the end of Mesozoic time, North America, powered by the slow roll of convection in the Earth's mantle, broke away from Europe. As the newborn Atlantic Basin widened, westward drift

of the continent and its collision with the East Pacific Plate caused new stresses in the crust. In what is known to geologists as the **Laramide Orogeny**, the present Rocky Mountains were born, rising east and north of the Plateau region but greatly influencing its development: They would provide headwaters for the great river that, with its tributaries, would carve and chisel the scenic Plateau country we know today. During the Laramide Orogeny, the Colorado Plateau began to rise.

The end of the Mesozoic Era may have been punctuated with great meteor showers, impacts of many asteroids, and a long night of airborne dust that led to the demise of more than half of the species of animals and plants then in existence.

Cenozoic Era. With the disappearance of the dinosaurs, mammals who had survived the great extinction spread and diversified. In Tertiary time, ancestors of today's horses, elephants, pigs, and camels roamed this region, feeding on the lush growth of a savannah environment. They in turn became prey to evolving members of the dog and cat families.

Uplift continued into early Cenozoic time. Breaking along old faults inherited from Precambrian rocks, segments of the Colorado Plateau jostled upward, slowly taking on their present elevations.

Uplift was accompanied, as always, by erosion, the stripping away of layer after layer of sedimentary rock. Because of the Plateau's alternating layers of hard and soft rock, most of the stripping occurs as cliffs of hard rock are undermined by the erosion of softer layers below. And because the rocks dip gently northward over most of the Plateau, most of the cliffs gradually retreat in that direction, a process still going on. North of Grand Canyon, successive lines of retreating cliffs demarcate the "Grand Staircase" of the Utah plateaus.

Today we can observe that streams tend to follow the lines of soft rock at the bases of such retreating cliffs, slowly migrating sideways as the cliffs retreat—a point not lost on geologists trying to decipher the history of the Plateau region in general and the Grand Canyon in particular. Most of the Mesozoic rocks have already been stripped away from the southern part of the Plateau.

As the Plateau developed, molten magma squeezed up along fissures and faults to create a number of domed laccoliths. Particularly around the margins of the Plateau, other magma burst through to the surface, building into several large composite or stratovolcanoes: San Francisco Mountain, Mount Trumbull, Mount Taylor, and the White Mountain volcanoes. Later, as the Colorado River and its tributaries were finally fashioning today's patterns of drainage, cinder cones and dark basalt lava flows added to the contrasts of Plateau country color. Some lava flows dammed the rivers, even the mighty Colorado, creating temporary lakes or bringing about changes in their courses.

The last eruption on the Plateau, that of Sunset Crater, occurred a little more than 900 years ago. In all likelihood, there will be other eruptions; the region is by no means "dead" volcanically.

The Ice Ages of Pleistocene time also helped to shape scenery here. A few mountain glaciers sculptured the highest peaks, carving scoop-shaped cirques and leaving rough piles of glacial debris at their lower ends. Rainy cycles in phase with continental glaciation farther north certainly added to the strength of the rivers carving the Plateau's many canyons.

The most recent changes in this region have been brought about by man, who has created dams and lakes where once there were canyons. Man, too, has cut wood and grazed cattle and sheep, too many of them, in this country, reducing vegetative cover and adding to the erodability of the land. In places he has gouged the earth for minerals, and a thickening pall of power-plant

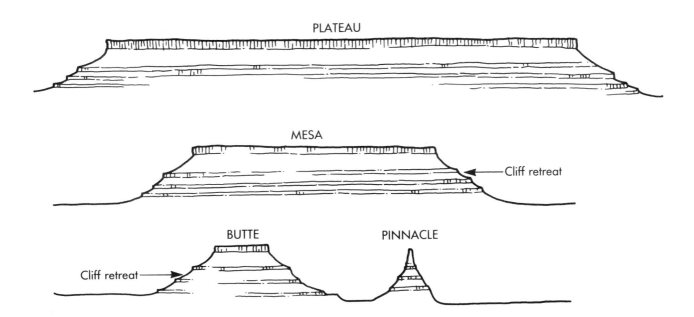

Most Plateau Country erosion involves the retreat of slopes and cliffs, as plateaus reduce to mesas and mesas reduce to buttes and pinnacles.

smoke reduces the clarity of the once pristine air.

In this book, as in others of this series, the parks and monuments are discussed in alphabetical order. In addition to the true Colorado Plateau area, this volume includes Black Canyon of the Gunnison National Monument, just east of what is usually defined as the Plateau.

Small maps distributed at entry gates by the National Park Service will help you find your way around in the parks. More detailed topographic maps and, for some parks, geologic maps are available at visitor centers.

IX. Other Reading

Most of the books and articles listed below cater to readers with little or no background in geology. The list does not include the many excellent and up-to-date textbooks now available at both high school and college levels.

Angier, Natalie, 1985. "Did Comets Kill the Dinosaurs?" *Time*, vol. 125, no. 18 (May 6), pp. 72-83.

Arizona Bureau of Mines and Mineral Resources. *Geologic Guidebooks to Highways of Arizona*. University of Arizona, Tucson.

Baars, D.L., 1972. *Red Rock Country*. Doubleday/Natural History Press, Garden City, New York.

Babcock, R.S.; Brown, E.H.; and others, 1974. *Geology of the Grand Canyon*. Museum of Northern Arizona, Flagstaff.

Barnes, F.A., 1978. *Canyon Country Geology for the Layman and Rockhound*. Wasatch Publishers, Inc., Salt Lake City.

Breed, William J., 1968. *The Age of Dinosaurs in Northern Arizona*. Museum of Northern Arizona, Flagstaff.

Bullard, F.M., 1962. *Volcanoes: in History, in Theory, in Eruption*. University of Texas Press, Austin.

Chronic, Halka, 1980. *Roadside Geology of Colorado*. Mountain Press, Missoula, Montana.

Chronic, Halka, 1983. *Roadside Geology of Arizona*. Mountain Press, Missoula, Montana.

Chronic, Halka, 1987. *Roadside Geology of New Mexico*. Mountain Press, Missoula, Montana.

Colbert, E.H. (editor), 1976. *Our Continent: a Natural History of North America*. National Geographic Society, Washington, D.C.

Colbert, E.H., 1983. *Dinosaurs of the Colorado Plateau*. Museum of Northern Arizona, Flagstaff.

Cowen, R., 1975. *History of Life*. McGraw-Hill Book Co., New York.

Crampton, G.C., 1983. *Standing Up Country—the Canyonlands of Utah and Arizona*. Peregrine Smith Books, Salt Lake City.

Decker, R. and B., 1981. *Volcanoes*. W.H. Freeman and Co., San Francisco.

Hintze, Lehi F., 1973. *Geologic History of Utah*. Brigham Young University Studies, no. 20, part 3.

King, Philip B., 1977. *The Evolution of North America* (revised). Princeton University Press, Princeton, New Jersey.

Kurten, B. 1972. *The Age of Mammals*. Columbia Press, New York.

Macdonald, G.A., 1972. *Volcanoes*. Prentice-Hall, Englewood Cliffs, New Jersey.

Nations, D., and Stump, E., 1981. *Geology of Arizona*. Kendall/Hunt, Dubuque, Iowa.

New Mexico Bureau of Mines and Mineral Resources. *Scenic Trips to the Geologic Past* (guidebooks to New Mexico highways). Socorro, New Mexico.

Powell, John Wesley, 1895 (reprinted 1961). *The Exploration of the Colorado River and its Canyons*. Dover Books, Mineola, New York.

Rigby, J. Keith, 1976. *Northern Colorado Plateau*. K/H Geology Field Guide Series, Kendall/Hunt, Dubuque, Iowa.

Rigby, J. Keith, 1977. *Southern Colorado Plateau*. K/H Geology Field Guide Series, Kendall/Hunt, Dubuque, Iowa.

Smiley, T.L.; Nations, J.D.; and others, 1984. *Landscapes of Arizona—the Geological Story*. University Press of America, Lanham, Maryland.

Stokes, W.L., 1969. *Scenes of the Plateau Lands and How They Came to Be*. Publishers Press, Salt Lake City.

Wilson, J.T., and others, 1972. *Continents Adrift. Readings from Scientific American*. W.H. Freeman and Co., San Francisco.

Coral-colored sandstone fins separated by narrow alleys set the stage for development of arches. Distant cliffs are in Cretaceous and Tertiary sedimentary rocks. National Park Service photo.

Part 2

THE NATIONAL PARKS AND MONUMENTS

Arches National Park

Established: 1929 as a national monument, 1971 as a national park
Size: 297 square kilometers (115 square miles)
Elevation: 1195 to 1723 meters (3900 to 5700 feet)
Address: c/o Canyonlands National Park, Moab, Utah 84532

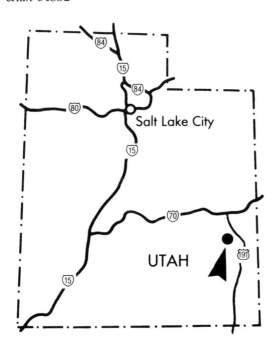

STAR FEATURES

• Stone arches—more than 200 of them discovered and documented—as well as windows, arch-shaped alcoves, and mazes of mini-canyons between tall fins of pink sandstone.

• Red and pink rocks deposited in river floodplains, on ancient dunes, and in near-shore lagoons.

• Unusual anticlines caused by the upward movement of underground salt, and valleys resulting from their collapse.

• Visitor center with geologic exhibits, a self-guided road tour, many trails (some with guide leaflets), guided tours in the summer describing geology, biology, and history.

See color pages for additional photographs.

SETTING THE STAGE

In eastern Utah, near the northern limits of the Colorado Plateau, flat-lying layers of sedimentary rock are corrugated by several northwest-trending anticlines. In Arches National Park, coral-colored sandstone of the Entrada Formation has responded to the collapse of two of these anticlines by splitting along parallel, northwest-trending, vertical joints. The joints widen as rain and snowmelt dissolve the calcium carbonate that binds the sand grains together. Soil develops along the joints, and plants find a foothold, furthering the dissolving process with products of their metabolism and decay. Seasonal freezing and thawing, rainwash, and wind widen the joints until the rocks stand as tall, slender fins.

As erosion continues, some parts of the rock fins weather faster than others, and alcoves and caves form as thin, curving sheets of rock fall away. Where such localized weathering attacks both sides of a fin—perhaps along a zone of weak-

The ultimate in arch formation, Landscape Arch is 88 meters (287 feet) long and 32 meters (106 feet) high. Yet it is only 3 meters (10 feet) thick at its narrowest point. Tad Nichols photo.

ness in the rock, perhaps where underlying mudstone channels moisture—caves may develop on both sides of a fin. Deepening, they ultimately join, creating a window or an arch.* Other arches form where solution works downward and then sideways from a pothole on the upper surface of a butte.

With time and further erosion, arches become slenderer and increasingly fragile. Eventually, unable to support their own weight, they collapse into piles of rubble. But arch formation is a continuous process, going on today: Fins are narrowing, new arches are forming, old ones are collapsing, and fallen rubble is being washed and blown away.

The secret to arch formation here is the pres-

ence of massive sandstone layers with abundant parallel vertical joints caused by bending across the Salt Valley and Cache Valley anticlines. The anticlines themselves are unusual: Cores of grayish and yellowish gypsum are exposed in valleys along their crests. These valleys commonly also contain tilted, irregular masses of rock younger than those exposed on their walls, suggesting that they were produced by collapse.

But what caused their collapse? The answer will surprise you. Salt. Well below the surface here are thick layers of salt, gypsum, and potash originally deposited as sea water evaporated in Pennsylvanian time. When deeply buried, these evaporite minerals become plastic, able to flow slowly like Silly Putty or glacial ice. Just as Silly Putty will flow out from under your fingers if you press on it steadily, the salt and gypsum tended to flow toward areas where overlying strata were slightly thinner and pressures therefore not quite as intense: in this region, along underground ridges that mark ancient northwest-trending faults.

---

* Natural bridges, as distinct from arches, span streams or stream courses. The term "window" is loosely used, but should be applied only to relatively small openings well above ground level.

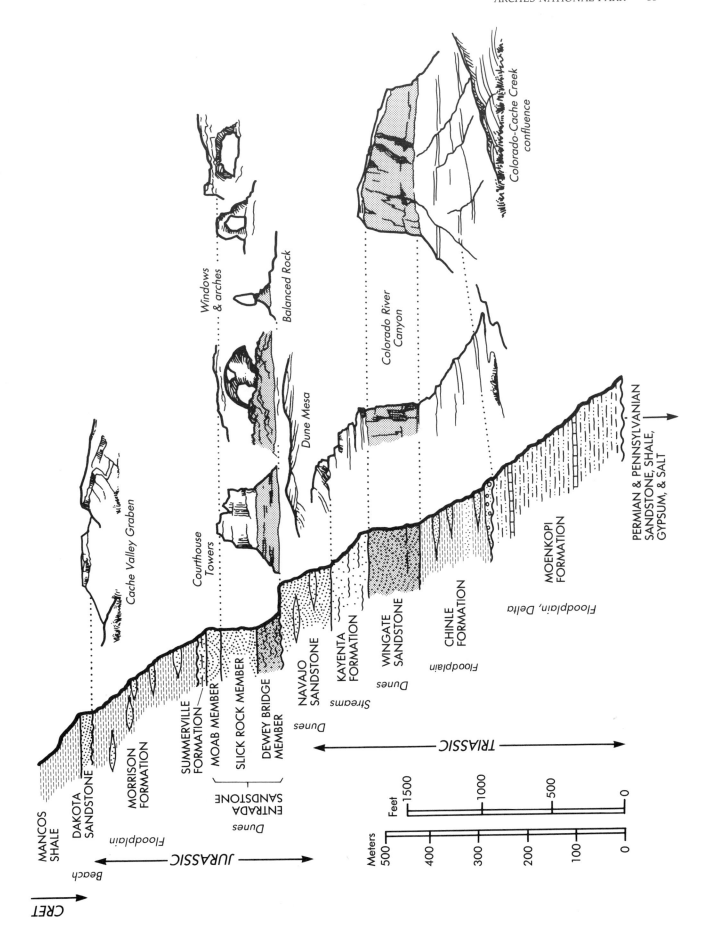

Colorado-Cache Creek confluence

Windows & arches

Balanced Rock

Colorado River Canyon

Dune Mesa

Cache Valley Graben

Courthouse Towers

MANCOS SHALE

DAKOTA SANDSTONE

Floodplain

MORRISON FORMATION

SUMMERVILLE FORMATION

MOAB MEMBER

SLICK ROCK MEMBER

DEWEY BRIDGE MEMBER

ENTRADA SANDSTONE

Dunes

NAVAJO SANDSTONE

Dunes

KAYENTA FORMATION

Streams

WINGATE SANDSTONE

Dunes

CHINLE FORMATION

Floodplain

MOENKOPI FORMATION

Floodplain, Delta

PERMIAN & PENNSYLVANIAN SANDSTONE, SHALE, GYPSUM, & SALT

Beach

CRET

JURASSIC

TRIASSIC

Feet

1500

1000

500

0

Meters

500

400

300

200

100

0

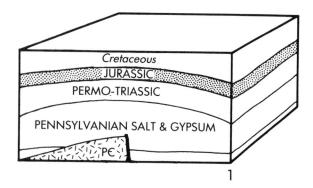

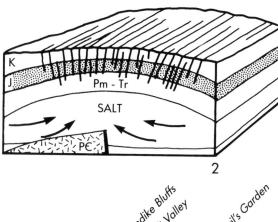

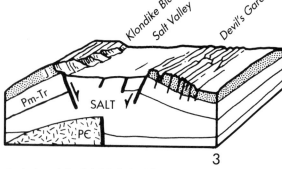

Evolution of the Salt Wash Anticline:

1. Thick layers of salt overlie an ancient fault. They are blanketed by hundreds of meters of Permian and Mesozoic rocks. Because of continued movement along the fault, sediments thin above it.

2. The weight of overlying rocks forces salt to flow upward toward the ancient fault, the area of least overburden, doming up and cracking overlying sedimentary rocks. Continued fault movement emphasizes the fold.

3. Removal of part of the salt by groundwater causes collapse of the anticline crest. Cretaceous and Jurassic rocks drop to form the valley floor, where gypsum remains as gray hills.

Less dense and more buoyant than overlying sedimentary rocks, salt and gypsum also pushed upward, exerting enough upward pressure to arch overlying rock layers into anticlines that inherited the northwest-southeast orientation of the ancient faults. Eventually most of the salt was dissolved away; only the gypsum now remains in the cores of the anticlines.

GEOLOGIC HISTORY

Precambrian Era. Although no Precambrian rocks are exposed within Arches National Park, their presence far below the surface has, as we have seen, influenced the geology and therefore the scenery here. The Precambrian rocks are probably like those exposed in the Uinta Mountains farther north and in Black Canyon of the Gunnison and Colorado National Monuments: gneiss, schist, and granite 2.4 to 1.5 billion years old, and weakly metamorphosed sedimentary rocks 1.2 to 0.6 billion years old. Late in Precambrian time these rocks were faulted into long northwest-trending ridges.

Paleozoic Era. During most of Paleozoic time, this region, along with practically all of the western United States, was low, almost flat, and often submerged by the sea. About 600 meters (2000 feet) of marine sedimentary rocks were deposited here. In Pennsylvanian time, a mountainous island—the Uncompahgre Highland—rose along the Colorado-Utah border, and in south central Utah a submarine platform began to restrict the easy flow of sea water. In a subsiding area between island and submarine platform, where circulation was periodically cut off, sea water evaporated, leaving thick deposits of salt and gypsum. Layers of black shale, usually taken as indicators of stagnant conditions, accumulated as well.

During Permian time these strata were covered with more marine sediments and with rock debris washing from the Uncompahgre Highland. And the salt and gypsum that accumulated in Pennsylvanian time began to flow toward the old Precambrian ridges, pushing up anticlines above them.

Mesozoic Era. For most of this era this region was above the sea. Triassic rocks record the retreat of the sea in their change upward from marine limestone to brilliant red shale and sandstone. These red sediments are the debris washed from the Uncompahgre Highland and deposited on river floodplains or a large delta, where they were later topped with dune deposits.

Late in Triassic time, mountains rising in central Utah blocked off moisture-bearing winds, and desert conditions prevailed. The thick, crossbedded, Triassic-Jurassic Navajo Sandstone, which forms finely ridged rock surfaces below the spectacular fins and arches in this park, is made of sand dunes of this great desert, as is the Entrada Formation of the fins and arches themselves. Above the dune sandstones are widespread sheets of rainbow-hued mudstone,

Figures like the Egyptian Queen result when hard sandstone layers are undermined by erosion of soft shale and mudstone like that in the foreground. The queen's head is thought to have been offset by an earthquake. National Park Service photo.

Double Arch formed, like other arches, as erosion of the crumply Dewey Bridge Member undermined the stronger sandstone above.

siltstone, conglomerate, and freshwater limestone deposited on river floodplains and in freshwater lakes.

Upward movement of Pennsylvanian salt continued sporadically through Triassic and early Jurassic time. Though many of the Triassic and Jurassic rocks thin and disappear near the salt anticlines, the uppermost Jurassic sediments once extended right across the anticlines.

As the Cretaceous Period began, seas again advanced, coming this time from the south and east. Few Cretaceous rocks are preserved in Arches National Park; in neighboring areas they show a cycle of shore sand, marine shale, and shore sand again. The upper sandstones are interlayered with river and delta deposits and coal beds that tell us of lagoons and swamps along the Cretaceous shore. The Cretaceous rock sequence can be seen to the northeast from high vantage points near Devils Garden: Book Cliffs, the lower band of cliffs curving across the Colorado-Utah border, display soft Cretaceous marine shale capped with the sandstone-coal-sandstone sequence. At the end of Cretaceous time, the sea withdrew.

Cenozoic Era. In late Cretaceous and early Tertiary time a wave of mountain building swept eastward from California and Nevada, buckling parts of eastern Utah and western Colorado and thrusting up the Rocky Mountains. The Arches area was caught between the new-formed San Rafael Swell of central Utah and a new Uncompahgre Uplift in much the same position as the earlier Uncompahgre Highland. The Arches area seems to have bowed downward, though not as sharply as regions farther west and north, where thick lake sediments accumulated. Some of the lake shales contain the beautifully preserved fish of Fossil Butte National Monument; some contain valuable deposits of oil shale. They now appear as Roan Cliffs, the upper range of cliffs visible to the north from high points within Arches National Park.

All during Paleozoic, Mesozoic, and Cenozoic time there were sporadic movements on the old Precambrian faults that underly this region and control the position of the salt anticlines. Movement of the salt and bowing up of the salt anticlines are believed to have occurred in two phases. The first phase began in Permian time and continued until Jurassic time—a very slow, perhaps pulselike process. A second phase in Tertiary time intensified the shape and size of the anticlines and led to their ultimate collapse.

During the Miocene and Pliocene Epochs, the whole region—Utah and Colorado and parts of adjacent states—rose 1500 meters (5000 feet) or so as a broad, almost indiscernible dome, bringing this region to its present elevation. Though many surface features remained nearly the same, the uplift increased rates of erosion. Rivers and streams, steepened and strengthened by uplift, bit into rock layers along their paths, carving the many splendid canyons of the Plateau country.

Meantime, with this region well above sea level, groundwater dissolved and removed the salt from the salt anticlines, flushing it away into the Colorado River, leaving only less soluble gypsum. As salt was removed from the anticlines, their crests collapsed, dropping in some cases many hundreds of feet. Each collapse carried down with it some of the younger rock that surfaced this area at the time. Each is now marked by long fault-edged valleys floored with gypsum and fragments of younger rocks, bordered with eroded bluffs, towers, and arches of sandstone younger than the gypsum but older than some of the rock fragments in the valleys.

Gradually, vast upland areas were swept clean of soft, easily eroded Tertiary deposits. Frost wedging, resulting from sharp temperature changes, loosened sand and rock fragments, which then were swept away by wind and rain. Though they brought no ice to this immediate area, the Ice Ages of Pleistocene time did bring cooler temperatures and probably greater precipitation, undoubtedly augmenting erosional processes. Streams and rivers became roaring floods, scouring ever deeper channels and sweeping ever more sand and silt from the highlands, bringing the scenery to what we see today.

BEHIND THE SCENES

Cache Valley Graben. One of the park's two collapsed salt anticlines, Cache Valley can be seen from viewpoints along the main park road or visited by the road to Wolfe Ranch. Bordering rock layers, which dip away from the valley, are the sides of the anticline. The crest of the anticline collapsed completely as its salt core was dissolved away, bringing down with it some of the rocks that covered the surface at the time of its collapse: Cretaceous Mancos Shale and Dakota Sandstone.

Bright green rock near the Wolfe cabin contains glauconite, a mineral thought to have been deposited in agitated water. Green marble-sized balls formed as concentric layers of glauconite were deposited on sand grains rolled to and fro by waves.

Surface water here is strongly flavored with salt and gypsum. Springs in recesses below the highest cliffs, which provided the Wolfe family with palatable drinking water, show that groundwater is moving slowly through the porous sandstone, especially just above an impermeable shale layer in the Entrada Sandstone.

Indian petroglyphs on the east side of Salt Wash, a little north of the Wolfe Ranch, were pecked through desert varnish that coats a sandstone slab, revealing the lighter rock below. Some were carved in historic time after horses were brought to the Americas by Spanish conquistadores. Even though the petroglyphs have been here for many hundreds of years, they show no trace of new desert varnish.

Colorado River Canyon. It's hard to get directly to the Colorado River from Arches National Park, but by going upriver from Moab you can enjoy a scenic drive and look across the river into the south edge of the park.

Courthouse Towers are isolated fins of Slick Rock Member sandstone undermined by erosion of the Dewey Bridge Member. National Park Service photo.

Rocks exposed along the river are older than those elsewhere in the park. The oldest are red sandstone and shale of the Moenkopi and Chinle Formations, of Triassic age. Above them rise high angular cliffs of Wingate Sandstone, a Triassic dune sandstone capped by thin, resistant ledges of the Kayenta Formation. Lighter colored, rounded cliffs along the skyline are Navajo Sandstone, dune-formed on a Jurassic desert. This rock forms the surface in the southern part of the park.

As you drive along the river, see if you can distinguish on the opposite bank the Courthouse syncline, Salt Wash anticline (where Salt Wash Creek enters the Colorado), and the highly faulted Cache Valley Anticline. All these structures extend south across the river but peter out near the La Sal Mountains.

Courthouse Towers. Courthouse Towers and the tall cliffs nearby are remnants of rock layers exposed also in the Windows and Devils Garden parts of the park, the two lower members of the Entrada Formation. Here they occur in a gentle syncline, its lowest point marked by Courthouse Wash. The valley is floored with Navajo Sandstone. Ledges and slopes at the bases of the cliffs and towers are in the Dewey Bridge Member of the Entrada Formation, recognizable by its wavy, crumpled bedding.

Here, as elsewhere in the park, joints caused by formation and collapse of a salt anticline dictate the major patterns of erosion. Minor patterns are governed by differences in rock composition and hardness. As they break away along vertical joints, resistant sandstone layers form cliffs and ledges. Shale and mudstone form slopes, benches, and recesses that in some cases undermine the massive sandstone cliffs. Where undermined, the sandstone may break away in arched alcoves.

Delicate Arch. The trail to Delicate Arch offers some of the most delightful, most geologic vistas in the park, with chances to examine the lowest two members of the Entrada Formation. The contorted Dewey Bridge Member, near the trail beyond the footbridge, in places contains clumps and nodules of agate. This unit weathers fairly readily along mudstone beds, and commonly undermines the massive dune sandstone above.

Farther along, the trail crosses smooth, barren slopes of the Slick Rock Member, pale pink sandstone swept by sand-dune cross-bedding. Furrows have developed along joints in this rock, and little potholes line up along the furrows. After rains, these potholes hold small pools, homes for tiny short-lived animals and plants whose metabolic by-products add acid to the water, speeding solution of the calcium carbonate that binds the sand grains of the rock together.

Delicate Arch and its surroundings display well the subtle differences of color and resistance to weathering that characterize the Slick Rock Member. Many small, deep cavities, carved by wind, line up along particularly easily eroded beds. Once these hollows are deep enough to hold a few sand grains, the wind is assured of a supply of tools, and whirling eddies bombard the hollows with sand, rounding and deepening them.

Devils Garden and Fiery Furnace. Red sandstone fins that result from erosion along parallel joints are thinner, higher, and more numerous here than in other parts of the park. The Devils Garden trail follows mini-canyons between the fins and finally climbs to the crest of one of them. Many arches are hidden in this rocky labyrinth, a number of them visible from this trail. All are different, and here you can pinpoint the factors that led to their development: erosion of the weak Dewey Bridge mudstone, closely spaced fracturing, weathering accelerated by the high moisture content of the rock, weathering along weak beds within the Slick Rock Member. Some arches are small and probably quite young. Others, like Landscape Arch (which may be the world's longest natural span), are about as large as they can get without collapsing.

Arch formation goes on today as it has for thousands of years: a sand grain or a pebble or a rock slab at a time. Skyline Arch, visible from the road, more than doubled its size in 1940 when a large block of sandstone fell from it, but most changes are not that sudden and dramatic. On many surfaces the rock peels off in flakes because temperature changes bring about small fractures parallel to the rock faces, and water freezing and thawing in the cracks flakes off curving sheets of rock. Many smoothly contoured surfaces are shaped in this manner, and many rounded arches are enlarged in this way.

Entrance Road. Geology around the visitor center is complicated by the Moab fault and the Moab Valley anticline, one of the northwest-trending anticlines of eastern Utah. The valley is lopsided: Rocks on its northeast side, behind the visitor center, are of very late Triassic and Jurassic age, younger than the Permian and Triassic strata across the valley. Relative displacement between the two sides of the valley is about 800 meters (2600 feet).

Leaving the visitor center, the road climbs through Jurassic strata—a good place to get acquainted with them, as exposures are top-notch and there are several turnouts for parking. The light buff or white Navajo Sandstone originated as sand dunes and interdune flats; it bears the long, sweeping laminations of the dunes and the silty horizontal beds of interdune deposits. Above it is gnarled red sandstone of the Dewey Bridge Member of the Entrada Sandstone. The Three Penguins are in the middle member of the Entrada Formation, the Slick Rock Member. This rock, too, is dune sand, and it is the layer in which are carved the arches of this national park. Several faults, with displacements up to about 15 meters (50 feet), are exposed on cliff faces near the road.

Beyond Courthouse Wash are many excellent exposures of the upper surface of the Navajo Sandstone, again showing the broad cross-bedding that characterizes dune sandstone.

Klondike Bluffs. These bluffs form the west limb of Salt Wash Anticline. They exhibit many sandstone fins and quite a few arches, among them Tower Arch with its stunning setting and lovely view. Along the trail to Tower Arch are displayed many characteristics of the rocks that form the fins and arches: the knobby, contorted slopes and "goblins" of the Dewey Bridge Member; the massive, cross-bedded, cliff-forming Slick Rock Member; and the thin-bedded white sandstone of the Moab Member, all parts of the Entrada Formation. Watch for examples of desert varnish, prominent diagonal layering of wind-deposited (and wind-eroded) sandstone, honeycomb weathering, and exfoliation, a form of weathering in which thin curving slabs peel off rock surfaces. Parallel joints control fin development here as on the northeast side of Salt Valley.

Park Avenue Trail. The contorted shale and mudstone of the Dewey Bridge Member and the hard, resistant, cliff-forming sandstone of the Slick Rock Member of the Entrada Formation are well exposed here. Notice the differences between ledge-forming sandstone and slope-forming shale and mudstone. Contorted bedding in the Dewey Bridge Member is thought to have originated before the mud of which it formed consolidated into rock—when it was soft enough to slump and slide. Some slumping may have been caused by the weight of overlying sediments, but the contortions do not extend up into the massive sandstone.

Desert varnish stains many cliffs above the trail.

Near Delicate Arch the fine laminations of wind-deposited sand festoon bare rock surfaces. Small holes are caused by wind erosion. Ray Strauss photo.

Salt Valley. At the heart of Arches National Park, Salt Valley is a collapse valley bordered by bluffs and cliffs of Jurassic sandstone. The road into it follows a creek bed through steep gray and yellow gypsum hills that are clues to the origin of the valley. When the crest of the Salt Wash Anticline collapsed, it carried down with it both Pennsylvanian gypsum and large fractured blocks of overlying strata.

Windows Section. Differential weathering of the lowest two members of the Entrada Formation, and the roles they play in arch formation, are particularly well demonstrated in this part of the park. The crumpled mudstone and sandstone of the Dewey Bridge Member almost invariably weather and erode more rapidly than the massive Slick Rock Member above. Balanced Rock is a fine example of this: The balanced boulder of Slick Rock sandstone perches on a pedestal of weaker Dewey Bridge mudstone. Both rise above a rolling surface of Navajo Sandstone, whose hard and soft layers weather in such a way as to emphasize its dune-derived cross-bedding.

Most of the arches here, including Double Arch, the North and South Windows, and Turret Arch, have also formed right at the Slick Rock-Dewey Bridge contact, where erosion of the mudstone undercuts overlying sandstone.

Pothole Arch, high up on the northeast side of Elephant Butte, developed in a different manner: A pothole on the upper surface of the butte deepened as rainwater pools, made acid by the metabolism of tiny aquatic plants and animals, dissolved the sandstone's calcium carbonate cement. At about the same time, a more or less horizontal recess formed in the wall of the butte along a weak layer in the sandstone. As erosion continued, cave and pothole met.

OTHER READING

Anonymous, no date. *The Guide to an Auto Tour of Arches National Park.* National Park Service.

Baars, D.L., and Molenaar, C.M., 1971. *Geology of Canyonlands and Cataract Canyon.* Four-Corners Geological Society, Durango, Colorado.

Hoffman, John F., 1985. *Arches National Park: an Illustrated Guide.* Western Recreational Publications, San Diego.

Johnson, David W., 1985. *Arches—The Story Behind the Scenery.* KC Publications, Inc., Las Vegas, Nevada.

Lohman, S.W., 1975. *The Geologic Story of Arches National Park.* U.S. Geological Survey Bulletin 1393.

Black Canyon of the Gunnison National Monument

Established: 1933
Size: 55 square kilometers (21 square miles)
Elevation: 1658 to 2609 meters (5440 to 8561 feet)
Address: P.O. Box 1648, Montrose, Colorado 81401

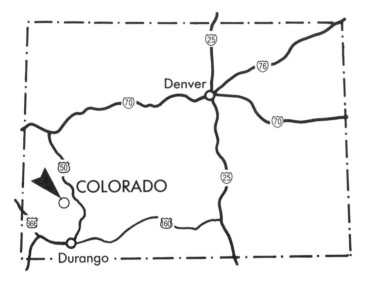

STAR FEATURES

• The deep, narrow canyon of the Gunnison River, carved in ancient rocks that form the foundations of the continent.
• An eventful canyon history involving mountain building, submergence and sedimentation, volcanism, and erosion.
• Visitor center, viewpoints, trails along the rim (some with trail leaflets), and, in summer, guided walks and evening programs.

SETTING THE STAGE

Black Canyon's name can be credited to its shaded depths, where the sun rarely shines, and to the dark rocks of its precipitous walls. In many places the canyon is deeper than it is wide. Here are some vital statistics:

Total length	.85 km (53 mi)
Length in national monument	19 km (12 mi)
Greatest depth	.739 m (2425 ft)
Average depth	.600 m (1968 ft)
Width of river at Narrows	.12 m (40 ft)
Width at rim above Narrows	400 m (1300 ft)

The walls of the canyon are composed of dark Precambrian gneiss and schist and lighter-colored granite. Veins of coarse-grained pegmatite slash obliquely across many parts of the canyon walls. Vertical joints form pathways for weathering and erosion, and as water seeps into them and freezes and thaws, contribute to the ruggedness of the canyon walls.

The Gunnison River, its headwaters high in Colorado's Rocky Mountains, carved this deep chasm with the torrents of its seasonal floods, when the untamed river battered rock with rock and swept debris downstream. Much of the carving dates to Pleistocene Ice Age time, when mountain glaciers released plentiful supplies of water and rock-fragment tools. In historic time the river has carried as much as 350 cubic meters (12,000 cubic feet) of water, sand, silt, pebbles, and boulders per second past any given point.

The river's steep gradient gave it the speed, power, and turbulence to cut downward faster than other processes could flare out its rims. Widening of the canyon is the work of frost action, rockfall, and landslide, with the river carrying away the fallen debris.

These processes continue today, but now upstream dams hold back floodwaters, so the river is unable to carry away the debris of passing years. Few tributaries enter the Black Canyon as it cuts across the high central core of the Gunnison Uplift, but those that do, meet the main canyon high up on its walls because they lack the cutting power of the larger river and cannot deepen their channels as effectively. In some parts of the canyon, lines of weakness such as joints, dikes, faults, and parallel mineral grains in schist provide avenues of drainage.

From viewpoints along the South Rim Drive, you can peer down into the somber chasm and see the hard crystalline rocks of its walls. It goes without saying that these ancient rocks are extremely resistant, extremely tough, capable of standing in near-vertical walls nearly twice as tall as man's tallest steel-reinforced structures. The rock units themselves are steeply tilted, in places almost vertical. Within the national monument they line up in broad bands that trend northeast-southwest, at right angles to most of the canyon, like books on a bookshelf waiting to be read by geologists and visitors.

Near Gunnison Point and the visitor center, dark layered gneiss appears in the canyon's shadowy depths. Some of this gneiss is grainy and very like quartzite; some is slaty or schistlike. Both types were baked and hardened, though never really remelted; they retain enough of their

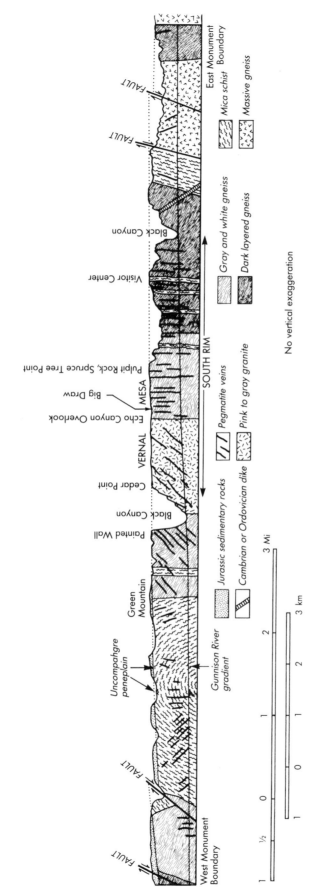

Ancient Precambrian rocks exposed in Black Canyon's walls are similar to those of Colorado National Monument and Grand Canyon's Inner Gorge. They range in age from 1.75 to 1.4 billion years old. Because of the river's winding course, this section crosses the river twice; the south rim shows in the center section.

former characteristics to show that they were derived from layers of sandstone, shale, and silt-stone. Some are still sandy in appearance, but with grains so tightly cemented that the rock breaks through the grains rather than around them—a characteristic of quartzite. Some contain scatterings of pebbles, for they were once con-glomerate.

Halfway between the monument headquarters and Pulpit Rock (Spruce Tree Point), where the South Rim Drive turns north, is some older metamorphic rock, gray and white gneiss marked with light and dark swirls. Large inclusions of granite are surrounded by darker ribbons of much finer-grained rock. Pure white quartz veins and light-colored pegmatite veins show up too, the latter with unusually large crystals of quartz, feld-spar, and white mica.

The origin of the gneiss is puzzling. Some of it looks as though it was squeezed as a semifluid porridge between layers of solid or slightly mushy older, darker rock. But in places the darker rock appears to have squeezed between masses of lighter rock, reversing this pattern!

Radiometric studies show that the gneiss is about 1.7 billion years old. Radiometric dating gives only the recrystallization age, however, and the original rocks of which it formed must have been older.

West of Echo Canyon Overlook, the rock type changes, and the South Rim Drive crosses a broad belt of coarse-grained granite. In this rock, flat-faced pink and white feldspar crystals and chunky, glassy quartz grains are large and easy to distinguish. (The pink color of the feldspar crys-tals results from long-term bombardment by radi-ation from radioactive minerals in the granite.) Smaller crystals of dark magnetite and hornblende give it a salt-and-pepper look. Little green crys-tals, hardly more than specks, are epidote.

The pinkish color carries through to the canyon walls, making this rock easy to recognize even from a distance. Its radiometric age is about 1.48 billion years, still well back in Precambrian time. It is, of course, older than the many coarse-crystaled pegmatite veins that cut through it. The Gunnison River runs right along the west bound-ary of the granite intrusion, at the foot of the Painted Wall, along a zone of weakness that seems to have guided its course. The edge of the intrusion continues southwestward and passes between High Point, at the end of the road, and Warner Point.

At the Painted Wall (the tallest cliff in Colorado) is more squiggly patterned gneiss, laced artisti-cally with pegmatite veins. The veins appear to

arch across the cliff, but they are really vertical sheets with a northeast-southwest trend almost parallel to the cliff face. The shape of the cliff face causes the archlike pattern.

Along the Painted Wall, vertical joints control erosion of deep parallel fissures. The remarkably flat upper surface of all the Precambrian rocks here is a surface that geologists call the Un-compahgre Peneplain. Back from the canyon rim, the peneplain is covered with flat-lying sedimen-tary rocks much younger and much weaker than the Precambrian rocks of the canyon itself.

Downcanyon from High Point and Warner Point is a fourth rock type: silvery mica schist that may well be the oldest rock in the canyon. It ap-pears in the center of an anticline, and walls most of the western part of the canyon.

GEOLOGIC HISTORY

Precambrian Era. The Black Canyon's story be-gins far back in Precambrian time, when preexist-ing sedimentary and perhaps volcanic rocks were heated and squeezed and ultimately recrystallized into schist and gneiss. Recrystallization is much more complete in the mica schist of the western part of the national monument than it is in the banded gneiss near the monument's headquar-ters, suggesting that metamorphism may have occurred in several stages. As the rocks cooled and contracted, fluid leftovers from the crystal-lization flowed into cracks and fissures, forming large-crystaled pegmatite veins. Throughout the crystallization process, crustal movements knead-ed the ancient dough, and eventually accordion-pleated it into tight anticlines and synclines with rock layers squeezed into vertical position.

After hundreds of millions of years of such alteration and deformation, magma rose from be-low, pushing aside some of the accordion-pleated older rocks or melting its way upward through them. Then the magma cooled and crystallized into the granite of the central part of the Black Canyon, with a new generation of pegmatite veins in its shrinkage cracks.

Late in Precambrian time, continent-wide ero-sion planed away the land, making of it, as the era closed, an almost featureless peneplain quite close to sea level.

Paleozoic and Mesozoic Eras. Soon after the Paleozoic Era opened, the beveled surface sank and marine sediments began to accumulate on it: thin sheets of limestone, shale, and sandstone, some entombing the shells of marine animals.

Black Canyon's spectacular chasm was carved by the swollen Gunnison River of Ice Age time. Though the canyon is 530 meters (1750 feet) deep, the river's present channel is in places only 12 meters (40 feet) wide.

The Painted Wall exposes the many light-colored veins that ornament the Precambrian gneiss. The horizontal surface at the top of the photograph represents several long, superimposed periods of erosion.

Late in the era, in Pennsylvanian time, several large islands were thrust above the sea, and streams and rivers stripped away the sedimentary rocks on their crests, once more laying bare the old Precambrian peneplain.

Later, in Triassic and Jurassic time, this surface was covered with shore sandstone and lagoon mud and clay, sediments that now appear as the sedimentary layers back from the northeast rim of the Black Canyon. Some layers contain limestone and gypsum, and suggest salty coastal ponds that dried up in an arid climate; some are of rainbow-hued mud deposited in river floodplains.

The Cretaceous Period brought a new incursion of the sea. Beach sand deposited on the retreating shore hardened with time to form the Dakota Sandstone, the resistant formation that now caps mesas near the national monument. As the sea deepened, the thick gray mud of the Mancos Shale accumulated on its floor, mud now exposed west of the Black Canyon in the fertile valleys of the Uncompahgre, Gunnison, and Colorado Rivers.

Cenozoic Era. Mountain-building forces struck again at the end of Mesozoic and beginning of Cenozoic time as the continent, having separated from Europe, drifted westward. This period of mountain building gave us most of the high ranges of the Rocky Mountains. The area around Black Canyon was broken once more along ancient faults, lifted, and tilted a little. New drainage patterns developed. At some time the ancestor of the Gunnison River was born, its course nearly the same as that of its present-day descendant.

Tertiary volcanism, however, gave the river a

hard time. As clouds of volcanic ash burst from volcanoes north and south of the ancestral Gunnison, its course was pushed southward, then northward, then southward again as the stream sought the low points between two volcanic ranges, the San Juan Mountains to the south and the West Elk Mountains to the north.

As volcanism continued, the land around the volcanic centers sagged, perhaps compensating for the vast outpourings of lava and volcanic ash. This sinking, too, affected the river, especially a ring-shaped subsidence that partly encircled the West Elk Mountains. Around 2 million years ago, as volcanism abated and finally died away, the Gunnison found its present course along the southern arc of this depressed, ring-shaped syncline.

The river cut easily through volcanic tuff and breccia and Mesozoic sedimentary rocks, soon deepening its channel. Once the channel was deepened, the river could no longer change its course, which lay by chance above the old uplift of hard Precambrian rocks. When eventually the river cut down to the uplift there was little it could do except continue to trench through the hard rock, carving out the canyon we see today. In this task it was helped by zones of weakness in the ancient rock: faults and joint planes and lines of contact between igneous and metamorphic rock.

No doubt downward cutting increased with glaciation, too, with abundant runoff in the Rockies, and with a plentiful supply of hard rock tools for the swollen river. Probably it was given a boost by changes in the course of the Colorado River, of which the Gunnison is a tributary, changes that led to steeper gradients throughout the Plateau region. To erode a canyon 600 meters (2000 feet) deep, cutting would have to proceed at an average rate of 33 cm (1 foot) every thousand years—only 1 centimeter (half an inch) in a human lifetime. Downward cutting is now drastically diminished by upstream dams.

OTHER READING

Dolson, John, 1982. *The Black Canyon of the Gunnison*. Pruett Publishing Company, Boulder, Colorado.

Hansen, W.R., 1971. *Geologic Map of the Black Canyon of the Gunnison River and Vicinity, Western Colorado*. U.S. Geological Survey Miscellaneous Geologic Investigations Map I-584.

Hansen, W.R., 1971. *The Black Canyon of the Gunnison, Today and Yesterday*. U.S. Geological Survey Bulletin 1191.

Bryce Canyon National Park and Cedar Breaks National Monument

Established: Bryce, 1923 as a national monument, 1924 as Utah National Park, 1928 as Bryce Canyon National Park; Cedar Breaks, 1933

Size: Bryce, 145 square kilometers (56 square miles); Cedar Breaks, 25 square kilometers (10 square miles)

Elevation: Bryce, 2018 to 2775 meters (6620 to 9105 feet); Cedar Breaks, 3155 meters (10,350 feet)

Address: Bryce, Bryce Canyon, Utah 84717; Cedar Breaks, P.O. Box 749, Cedar City, Utah 84720

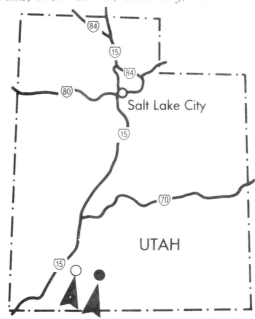

STAR FEATURES

• Two gigantic amphitheaters ornately decorated with colorful turrets, pinnacles, and free-standing walls carved as streams erode headward into two of the High Plateaus of Utah: the Paunsaugunt and Markagunt Plateaus.

• North-south faults that separate the plateau segments.

• Soft pink sedimentary rocks deposited in a Paleocene lake.

• Drives along the forested rims of the two plateaus, with viewpoints overlooking the eroded landscape.

• Many trails, some with self-guide leaflets, along the canyon rims. At Bryce, trails also lead down among the carved pink rocks.

• Visitor centers, museums, and, in summer, guided walks and evening programs.

See color pages for additional photographs.

Bryce Canyon's sculptured walls reveal the layered lake deposits of the Claron Formation. Here, hoodoos are beginning to form. Vertical mud stalactites (far right) develop as rain washes mud from the hoodoos. Ray Strauss photo.

SETTING THE STAGE

Along the east margin of the Paunsaugunt Plateau and the southwest tip of the Markagunt Plateau, mischievous nature has carved fanciful fairylands of elaborate cliffs and pinnacled ridges separated by deep, steep alleyways. The intricate, many-branched mazes are sculptured in soft pink rock of the Claron Formation (formerly assigned to the Wasatch Formation), made up almost entirely of limy Paleocene lake deposits that range from silty and sandy limestone to limy siltstone and sandstone, with liberal doses of clay and volcanic ash. The pink color of these rocks comes from minute amounts of iron oxides, mostly tiny grains of hematite. At Cedar Breaks, the entire formation is considerably thicker, sandier, and siltier than at Bryce.

Below the lake deposits are older rocks—tan or gray Cretaceous sandstone—equally susceptible to erosion but tending to form ridged hills rather than steep gullies and ridges.

At both sites the lake limestone is cut by faults. At Bryce, the Sevier Fault on the west, bordering the Sevier Valley, and the Paunsaugunt Fault on the east, define the Paunsaugunt Plateau. At Cedar Breaks, the Hurricane Fault edges the

Two major faults define the Paunsaugunt Plateau. On both, the east side is lifted relative to the west side.

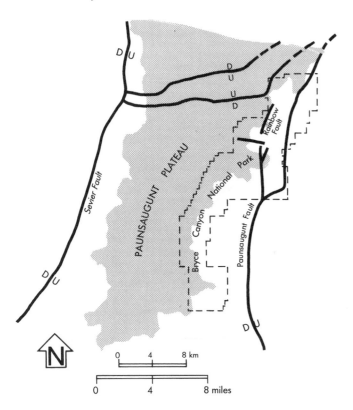

Markagunt Plateau. Along all three of these faults, the east side is uplifted relative to the west side.

Near Bryce the Sevier Fault, entirely outside the park, is exposed near the mouth of Red Canyon, where the pink rocks are pushed upward some 300 meters (900 feet), and abut basalt dated at 500,000 years old, the fault movement obviously being more recent than that. The Paunsaugunt Fault, along the east side of the plateau, is partly in the park. On it, displacement exceeds 200 meters (650 feet).

Several lesser faults, with offsets ranging from a few centimeters to several meters, parallel or cut across these two main faults. The Fairyland Fault in the northern part of Bryce is of special interest because it offsets all but the uppermost rock layers, those that form the caprock of the Paunsaugunt Plateau; apparently its movement took place before the caprock was deposited.

At Cedar Breaks, the Markagunt Plateau drops off abruptly along the Hurricane Fault, which effectively separates the Colorado Plateau as a whole from the Basin and Range region to the southwest.

At both Bryce and Cedar Breaks, many vertical joints slice through the rock as well, most of them due to wrenching and breaking during uplift of their respective plateaus. Both faults and joints play important roles in directing erosion of the two dramatic escarpments.

For erosion is king here: erosion governed by plateau uplift, by the downward erosion of the Colorado River and its tributaries, by the relative weakness and porosity of the limestone layers, by joints and faults, and by climate. And what erosion! Towers and turrets, narrow scalloped walls, and thousands of ornate and fanciful figures populate the descending ramparts, delicately touched with color that varies in intensity from place to place as well as from hour to hour, day to day, and season to season.

The primary agent of this erosion is water, in the form of rain, frost, snowmelt, and runoff. The work of water is also abetted by wind and gravity. Erosion initially followed joints and faults, which allowed passage of groundwater and solution of the calcareous (limy) cement binding the siltstone and sandstone particles together. Gradually the joints and faults widened into gullies. When tributary gullies developed, they either followed the joint pattern of the rock, or sprang from fortuitous headward erosion into the rims of earlier gullies.

As erosion claws at the edges of the Paunsaugunt and Markagunt Plateaus, the sculptured cliffs gradually retreat. In steep gullies, runoff from summer thunderstorms or spring snowmelt

Hoodoos on the right skyline are remnants of a wall like that in the center of the photograph. Hoodoo development is most marked in the middle of the Claron Formation. Ray Strauss photo.

sweeps loose debris downward. Debris sliding through large gullies concentrates along their edges, where bits of rock and sand fall from the gully walls. As they move—either washed by water or sliding landslide fashion—they carve smaller gullies and leave grooves or scour marks in the rock walls, marks that show past positions of the steep gully floors. Undermining resistant rock, their movement helps to maintain the sheerness of gully walls.

Especially at Cedar Breaks, which is higher and colder than Bryce, water freezing in cracks and crevices pries rock blocks away from the upper parts of the cliffs. In the southern part of Bryce, where there are several prominent limestone and sandstone ledges, similar frost wedging loosens large rock blocks, which tumble from the cliff faces to slopes below. In both park areas, frost also works at a lesser scale, separating smaller blocks and even individual sand and silt particles from the porous, water-absorbing, loosely ce-

mented limestone. As broken rock material falls, talus slopes and cones form below the cliffs; then frost heaving takes over, moving the talus fragments gradually downslope. Both frost wedging and frost heaving depend on day-night temperature variations, which are most extreme (with overnight freezing and daytime thawing some 200 to 300 times per year) on south-facing slopes near the rim.

Horizontal corrugations in the walls of sculptured alleyways and in the cliffs and turrets are due to natural variations in the composition of the lake sediments. Limestone and sandstone in general are more resistant than layers containing large proportions of silt and clay. Dolomite—like limestone but with added magnesium carbonate—is more resistant still. Dolomite and well-cemented sandstone cap much of the rim and many prominent ledges, walls, and hoodoos. In many places, the clayey residue from solution of calcium carbonate coats walls and hoodoos like stucco, concealing variation between the layers.

None of the erosion would be possible without differences in elevation caused by faulting. Normally, uplift sets the stage for erosion, as can be seen clearly at Cedar Breaks. But oddly, the Bryce portion of the landscape is on the *low* side of the controlling Paunsaugunt Fault. Land east of the fault, once raised higher, has been eroded even more severely, until the pink limestones and siltstones of the Claron Formation are gone entirely. Northeast of Bryce, where they are protected by lava caprock, they still remain as the Pink Cliffs of the Aquarius Plateau.

Well south of Bryce and Cedar Breaks, the Colorado River dictates how much erosion *can* take place. As the river carves canyons into older, harder rocks, its tributaries—among them the Paria and Virgin Rivers and *their* tributaries—

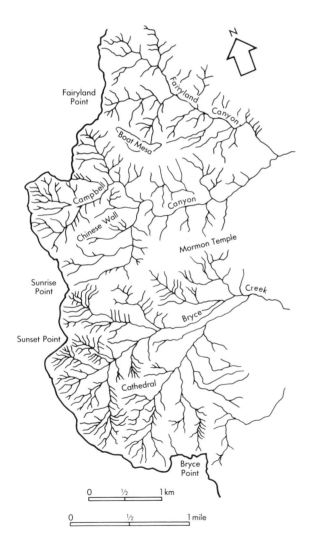

Eroding headward, Fairyland, Campbell, and Bryce Creeks have branched and rebranched in a dendritic drainage pattern that locally, where gullies are parallel, reflects joint sets in the rocks.

keep pace, cutting through young, soft rock, working headward, shaping semicircular amphitheaters. The erosion pattern is very much one of slope and cliff retreat, a pattern often seen in horizontal or nearly horizontal rock layers.

The rate at which the Bryce and Cedar Breaks amphitheaters are eroding headward has been determined by tree ring studies. Many of the trees along the canyon rims now stand on tiptoe, their roots bared by continuing erosion of the edge of the canyon rim. When compared with a tree ring "yardstick" developed by archaeologists for dating Southwestern ruins, slender cores taken from these trees (with no harm to the trees) show that different parts of the rims recede 24 to 120 centimeters (9 to 48 inches) per century, a rapid change as geologic processes go. As the cliffs recede, small changes take place in the varied forms and figures of both amphitheaters. Some walls, some hoodoos fall; others, in unending succession, take their place.

Bryce Canyon National Park is on the ridge between two drainages, that of the Paria River and that of the Sevier River. In dramatic contrast with its eastern slope, the western edge of the Paunsaugunt Plateau shows few of the intricately carved cliffs and pinnacles displayed in Bryce. There, summer temperatures are lower, snow insulates the ground in winter, moisture retention is higher, and vegetation is dense. Under these conditions erosion is much less severe. The Paria River and its branching tributaries erode the east side of the Paunsaugunt Plateau faster than the Sevier River and *its* tributaries erode the west side. Eventually the Paria tributaries will do away with the plateau altogether, breaking through the ridge and "capturing" the Sevier River's drainage.

GEOLOGIC HISTORY

Mesozoic and Cenozoic Eras. Paleozoic and early Mesozoic history is not represented in the Bryce and Cedar Breaks areas, but is similar to that of the rest of the Plateau. Thick layers of Cretaceous marine and near-shore sedimentary rocks underlie the Claron Formation. As North America drew away from Europe at the end of Mesozoic time, the continent rose slightly, and the widespread sea in which the rocks were deposited withdrew. In the continental interior rose two ranges destined to play parts in the Bryce–Cedar Breaks story: the Rocky Mountains in Wyoming and Colorado, and the Sevier Mountains in Utah. In landlocked basins between the two ranges, large lakes received runoff and finely pulverized rock debris from both.

One of these inland lakes stretched diagonally across Utah from northeast to southwest. At its greatest extent it measured about 400 kilometers (250 miles) long and nearly 120 kilometers (75 miles) wide—slightly larger than Lake Erie. In the lake, in Paleocene time, silty limestone and mudstone lake deposits accumulated to depths as great as 300 meters (1000 feet), to become, later, the Claron Formation.

Starting about 35 million years ago, in Oligocene time, and continuing well into Miocene time, volcanoes erupted nearby, creating, among other features, the lava flows that can now be seen atop the Aquarius Plateau. The Colorado Plateau as a whole began to rise, with volcanism and intrusions marking its margins. By late Miocene time, around 12 to 15 million years ago, the various segments of the Colorado Plateau had separated and moved up or down by different amounts.

Uplift was of course accompanied by erosion. As the Colorado River became the master stream of the area, headward erosion along its tributaries gradually created features we see today—the canyons, the lines of cliffs, the desert badlands that characterize this region. One of these tributaries, the Paria River, worked its way headward, branching upstream into many smaller tributaries. Carving out narrow canyons and even narrower gullies, these tributaries came together on the broad, gently sloping floor of Paria Amphitheater. Stable for a time, this broad, semicircular amphitheater was only fairly recently (probably less than 500,000 years ago) dissected by the very streams that had shaped it, leaving remnants of its old floor as sloping mesas within the present wide valley. At Cedar Breaks, Ashdown Creek has shaped the land in much the same way, eroding headward into the edge of the Markagunt Plateau.

BEHIND THE SCENES

Cedar Breaks. Headward erosion of Ashdown and Rattle Creeks and their small tributaries is responsible for this gigantic amphitheater. The colorful, intricately carved cliffs have the same fairyland appearance as Bryce Canyon.

The Claron Formation is considerably thicker and more colorful here than at Bryce. We are near one of its main sources: the Sevier Mountains of early Tertiary time. The formation is also considerably sandier and siltier, and contains less limestone and dolomite and more volcanic ash, than its counterpart in Bryce. Grayish rocks underlying the pink strata are Cretaceous; many of them belong to the Kaiparowits Formation, made up of deposits of an older freshwater lake.

Southwest of Cedar Breaks, Brian Head, 3466 meters (11,307 feet) high, is capped with lava flows and volcanic ash that erupted in Eocene time. Quaternary lava flows—some of them quite fresh in appearance, as if they had erupted only yesterday—surface parts of the Markagunt Plateau, in places damming small ponds and lakes.

There are quite a few bristlecone pines at Cedar Breaks; see the comment under Bryce Point, below.

Bryce Canyon viewpoints and trails are described below in north-to-south order, as they appear on entering the park by car.

Fairyland Viewpoint. The northernmost viewpoint in the park reveals the Pink Cliffs, carved in the Claron Formation lake sediments, and provides a good view of lava-capped Aquarius Plateau to the north, upfaulted relative to the Bryce area. Though this part of Bryce is not as deeply eroded as portions farther south, and slopes are not as steep, several interesting features are apparent. The Fairyland Fault transects Boat Mesa

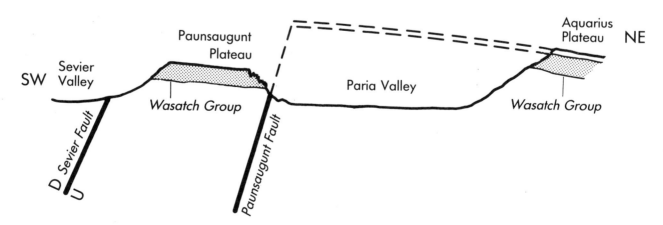

The Paria Amphitheater was carved in the uplifted block of the Aquarius Plateau. More Pink Cliffs appear along the edge of the higher plateau, raised well above the level of their counterparts at Bryce.

Bands of hoodoos and finlike walls reflect the systematic layering of the lake deposits. Some layers erode into hoodoos; others form ledges or slopes. Ray Strauss photo.

just to the south, offsetting some of the pink limestone and siltstone layers. Since the rimrock is not offset, movement along this fault must have preceded its deposition. On the Sinking Ship to the southeast, the Paunsaugunt Fault brings gray-brown Cretaceous rocks hard up against pink limestones of the Claron Formation.

Fairyland Trail (13 kilometers, or 8 miles). Going down Fairyland Canyon, winding around the base of Boat Mesa, this trail offers good views of the tilted fault block of the Sinking Ship. The Paunsaugunt Fault cuts across the low "stern" of the ship, with the pink Claron lakebeds, tilted by drag along the fault, abutting gray and tan sandstone of the Wahweap and Straight Cliffs Formations.

A short side trail visits Tower Bridge. The main trail then climbs toward the Chinese Wall and returns to the rim near Sunrise Point. Good examples of differential erosion of hard and soft

rock, and of the gullying process in general, appear along this part of the route, along with many views of other scenic and geologic features.

Sunrise Point. This viewpoint is on the dividing line between the less sharply eroded amphitheater formed by Campbell Creek and its tributaries and, south of it, the steeper, more ornate, more spectacular amphitheater of Bryce Creek and its tributaries. Several caprock layers control erosion here, with deep gully development initiated by erosion along joints.

Queens Garden Trail (2.4 kilometers, or 1.5 miles) descends from Sunrise Point into an exceptionally sharply dissected part of the Bryce amphitheater, with steep cliffs and many pinnacles. Headward erosion of tributaries of Bryce Creek is at its most spectacular here. Gradients are very steep, and gully walls, protected by hard caprock and scoured by downward movement of debris in the gullies, are essentially vertical.

The Sinking Ship is a tilted fault block. Pink cliffs of the Aquarius Plateau show on the skyline. Ray Strauss photo.

Sunset Point. Looking northeastward across the Paria Amphitheater to the Aquarius Plateau, southeastward toward Bryce Point, and downward into both the Queens Garden and the intersecting alleyways of the Silent City, views from Sunset Point are the most dramatic in the park. Resistant rock layers stand out above less resistant ones. Gullying is rapid, with headward erosion cutting back the rimrock and steep slopes below it, a process still going on. The amphitheater shape of the Silent City, formed by headward erosion of gullies tributary to Bryce Creek, shows up well between here and Bryce Point.

Inspiration Point provides another look down into the Silent City with its steep alleyways and light-reflecting walls. Between here and Bryce Point many large blocks of rimrock have fallen

from the rim, pried loose along joints by frost wedging or undermined by erosion of softer rocks below.

Bryce Point. This high point is on an upraised block between two faults: the Bryce Point Fault (a reverse fault, rare in the Plateau region) and the Peekaboo Fault. The gnarled trees on Bryce Point (and also on Inspiration Point, along the rim of the Queens Garden, and near Rainbow and Yovimpa Points) are ancient bristlecone pines, the species that in California includes the oldest living things in the world—more than 4000 years old.

Peekaboo Loop Trail (5.5 kilometers, or 3.5 miles). Dropping from Bryce Point into the Bryce Amphitheater, this trail covers some of the most dramatic ground in the park. If transportation is available, take this trail to its junction with the

Queens Garden Trail, then ascend to the rim at Sunset Point. Along the trail are abundant examples of Bryce's erosional features.

Paria View. Looking out over the Paria Amphitheater, this view shows especially well the flat-topped older mesas in the distance. These mesas are remnants of the floor of an earlier Paria Amphitheater, dissected by a new cycle of downward and headward erosion. Other features visible from this overlook include foreground foothills shaped in gray Cretaceous sandstone and shale, lava-capped Aquarius Plateau to the northwest, and in clear weather the distant Henry Mountains, an eroded laccolith.

Natural Bridge. Formed where erosion by wind and rain attacked both sides of a narrow limestone fin or wall, this hole in the rock is a natural arch, not a true natural bridge, which would involve stream erosion. Note that it formed along a joint, where percolating water weakened the rock by dissolving its calcium carbonate cement and, freezing and thawing, loosened its sand grains.

Rainbow and Yovimpa Points. At the southern end of the park road, these two points are unsurpassed for views to the south and southeast. The wide dome of the Kaibab Plateau, north of Grand Canyon, can be seen from Yovimpa Point. At both these viewpoints, erosion takes on a different character from that displayed farther north. Blocks of rimrock, wedged apart by frost and undermined by erosion of the soft sedimentary rocks below, have broken away and fallen. Accumulating on the talus apron below the cliffs, they move gradually downslope because of frost heaving. Frost action is a particularly potent factor here in the highest part of the park. Gray foothills below Yovimpa Point show that erosion reaches well down into Cretaceous rocks.

OTHER READING

Bezy, John, 1980. *Bryce Canyon: The Story behind the Scenery.* KC Publications, Inc., Las Vegas, Nevada.

Breed, W.J., 1983. *Geologic Cross Section of the Cedar Breaks–Zion–Grand Canyon Region.* Zion Natural History Association, Springdale, Utah.

Gregory, H.E., 1951. *The Geology and Geography of the Paunsaugunt Region, Utah.* U.S. Geological Survey Professional Paper 226.

Lindquist, Robert C., 1977. *The Geology of Bryce Canyon National Park.* Paragon Press, Salt Lake City.

Canyon de Chelly National Monument

Established: 1931
Size: 337 square kilometers (130 square miles)
Elevation: 1689 meters (5540 feet) at visitor center
Address: Box 588, Chinle, Arizona 86503

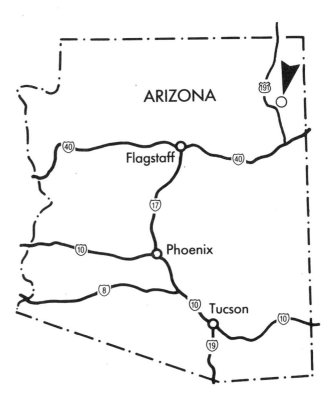

STAR FEATURES

• Two scenic canyons with soaring walls of pink sandstone.
• A major anticline, the Defiance Uplift, that brings these rocks to our view.
• Natural rock alcoves sheltering homes of the prehistoric people who once made these canyons their home.
• Two mountain-fed streams that water farms of present-day Navajo Indians.
• Rim drives, trails, canyon tours (with Navajo guides), visitor center/museum, and evening programs.

SETTING THE STAGE

The Defiance Uplift, an oval-shaped dome 150 kilometers (100 miles) long and 50 to 65 kilometers (30 to 40 miles) wide, is steeply inclined on its eastern flank but slopes more gently on its western side, where this national monument is located. Two thousand meters (6000 feet) higher than

The free-standing pinnacle of Spider Woman Rock (right) is an erosional remnant of De Chelly Sandstone remarkably similar to the not-yet-isolated rock to the left. Ray Strauss photo.

most of the Mesozoic sedimentary strata that surround it, the uplift has eroded down and into older rock layers, most notably the Permian De Chelly Sandstone, the pink sandstone exposed in the towering walls of Canyon del Muerto and Canyon de Chelly, both within this national monument.

In this area the Defiance Uplift is surfaced with hard, resistant Shinarump Conglomerate, the lowest unit of the Triassic Chinle Formation. Lying directly on the De Chelly Sandstone, this resistant unit is a mixture of coarse sand and rounded pebbles of Precambrian rock and of particularly resistant Paleozoic rock such as chert, which is abundant as nodules in Paleozoic limestone. Short, straight, stream-type cross-bedding in the Shinarump Conglomerate shows up well when seen at a distance across the canyon from rim viewpoints.

The exposed upper surface of the Shinarump Conglomerate, scoured by wind and rain, is dotted with shallow potholes. After rains, when water stands in the potholes, tiny plants and animals populate these mini-ponds, living out their lives in the short days before their pools dry up. These organisms secrete weak acids, products of their metabolism, that join with the slight natural acidity of rainwater to attack the calcium carbonate cement that holds Shinarump sand grains together. When the ponds evaporate, wind blows away the loosened grains, deepening the potholes. Wind winnows other parts of the Shinarump surface, too, carrying away dust and sand but leaving the heavier Shinarump pebbles as close-packed desert pavement.

From many of the rim viewpoints, the sharp contact between the Shinarump Conglomerate and the pale, peach-colored De Chelly Sandstone below it is clearly exposed, a contact that represents 60 million years of erosion or nondeposition. Here and there along the rim, stream channels filled with cross-bedded sandstone and conglomerate cut down into the De Chelly Sandstone.

Canyon walls expose the massive Permian sandstone. Cross-bedding marks the De Chelly Sandstone with sweeping laminae and tells us of its wind-deposited origin. The sand grains are uniform in size, well rounded, and lightly frosted or pitted from collisions with other wind-borne sand grains. On the cliff walls, horizontal laminae intercept or bevel the sloping ones, marking transient interdune surfaces like those known amid modern dunes. On some of these flat surfaces, thin, flat-lying layers of silt accumulated, interdune deposits that appear and disappear over relatively short distances.

Undermined by stream erosion, this massive rock weakens along vertical joints, and ultimately giant slabs crash to the floor of the canyon. Undermining was doubtless much more forceful in the past, when Ice Age climates brought lower temperatures and probably increases in precipitation. The long, frigid winters of the Ice Ages undoubtedly meant increased day-to-night and season-to-season temperature changes that repeatedly froze water in the joints, helping to pry slabs away from the walls.

The cross-bedding forms the sloping ceilings of many recesses and alcoves in Canyon de Chelly and Canyon del Muerto, particularly in the deeper eastern parts of the canyons. Localized concentration of moisture above silty, clayey, less permeable interdune deposits accounts for the erosion of most of these caves. As frost worked on rock already weakened by solution of its calcium carbonate cementing material, thin slabs and sheets of rock, and occasionally large blocks, fell away and deepened some of the recesses until they were large enough to be used as living space by "the ancient ones," the Anasazi peoples who inhabited this region 700 to 1500 years ago.

The walls of Canyon de Chelly and Canyon del Muerto are streaked with shiny stains of manganese and iron oxides—desert varnish—and with dull ribbons of carbonaceous plant material washed from overlying soils or formed in place as algae, lichens, and mosses grew on frequently dampened surfaces. The smooth desert-varnish veneer of some rock faces tempted early inhabitants, as well as later Indians, to peck petroglyphs (rock pictures) through the dark sheen and into the pale rock behind.

The massive walls of Canyon de Chelly and Canyon del Muerto become higher and higher eastward as the De Chelly Sandstone rises toward the summit of the Defiance Plateau.

GEOLOGIC HISTORY

Paleozoic Era. The Defiance Uplift seems to be a very old structure, one that more than once rose above the sea. On it, Permian rocks rest right on Precambrian rocks. It seems likely that most of the older Paleozoic rocks that occur elsewhere in the Plateau region were deposited here, but that they eroded off as the area lifted above sea level in Devonian and Pennsylvanian time. During Permian time, though the area still tended to remain high relative to its surroundings, deposition reached across the arch of the uplift again. The deposits—some marine, some continental—demonstrate the nearness of the Permian shore.

Towering walls of De Chelly Sandstone reveal sloping cross-bedding of Permian sand dunes. Carbonaceous plant material derived from lichens adds long streaks above White House Ruin.

Hunters on horseback post-date Spanish introduction of horses. These petroglyphs were pecked through a layer of desert varnish.

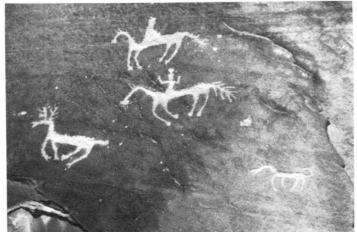

Cross-beds in the De Chelly Sandstone almost all slope southwestward, showing that the prevailing winds that formed them blew predominantly from today's northeast. The greatest present-day deserts, the Sahara and Arabian deserts and the dry interior of Australia, lie 20 degrees to 30 degrees north and south of the equator, in zones where the easterly trade winds of the tropics meet the westerlies of temperate belts. The deserts of Permian time, which extended north and east beyond Arizona's boundaries, lay in a similar belt relative to the position of the Permian equator, which ran obliquely across what is now North America.

Mesozoic Era. In early Triassic time, mountains rose in central and southern Arizona, high ranges that shed coarse sand and gravel northward across a broad, flat coastal plain. A thin layer of debris from these mountains is today the Shinarump Conglomerate, hard rimrock of Canyon de Chelly and Canyon del Muerto. Older highlands in northwestern New Mexico and southwestern Colorado may also have supplied rock debris to this area.

In most of northeastern Arizona the Shinarump Conglomerate is underlain by Triassic red-brown shales and mudstone—the "redbeds" of the Moenkopi Formation. Here on the Defiance Uplift, however, no such redbeds exist. Their absence, as well as the absence of many Paleozoic strata that we might expect below them, is evidence that the Defiance Uplift has been present, off and on, for hundreds of millions of years.

Later in Triassic time the nature of the deposits changed. Volcanoes somewhere to windward contributed great quantities of volcanic ash, the raw material of most of the Chinle Formation. Decomposing into several different clay minerals of a group known collectively as bentonite, the volcanic ash today is responsible for the many-hued badlands of the Painted Desert. Some of it seems to have been stream deposited, probably carried by floods far greater than but similar to those that followed the 1980 Mount St. Helens eruption. The floods may also have brought with them, as did the Mount St. Helens floods, the battered tree trunks now preserved in Petrified Forest National Park.

Late Triassic and early Jurassic time saw the accumulation of great seas of windblown sand across a broad desert that must have resembled the Sahara and Arabian Deserts of today. Many hundreds of feet of dune sand—now the Navajo Sandstone—accumulated in deserts that stretched from southern Nevada to Wyoming. Then, after several rounds of river and stream and volcanic ash deposits, the land was once more beveled by erosion.

During Cretaceous time a wide sag developed across a region that extended far north and east of Arizona. An arm of a southern sea crept in across this part of the continent, first depositing a thin layer of near-shore sands—the Dakota Sandstone—and then a thick layer of dark gray marine shale—the Mancos Shale. As the sea retreated, these deposits were capped by near-shore, lagoon, and beach deposits, the sandstone, shale, and coal sequence of the Mesaverde Group.

None of these Jurassic and Cretaceous rocks appear in Canyon de Chelly National Monument, but they can be seen across Chinle Valley in the sloping walls and hard rock cap of Black Mesa.

Cenozoic Era. Within the national monument no record remains for the Cenozoic Era. But not far to the south, flat-lying lake- and stream-deposited sandstone, conglomerate, and limestone layers of the Bidahochi Formation tell of a large Miocene-Pliocene lake that along its northeastern edge slowly filled with sand and gravel washed off the Defiance Uplift. The Chuska Sandstone, high up on the Chuska Mountains east of Canyon de Chelly, is of about the same age; its lower part was clearly deposited by streams, and its upper part as dunes.

Volcanism played a role in establishing the Chuska Mountains: Lava flows and fragmented and rewelded volcanic breccia cap this range and shield it from the erosion that has bared much of the surrounding country. Though the Chuskas were untouched by glaciation, the heavier rainfall and snowfall of Pleistocene time did facilitate slumps and landslides in these mountains, and doubtless affected rates of downcutting in Canyon del Muerto and Canyon de Chelly.

OTHER READING

Supplee, C., and Anderson, D. and B., 1971. *Canyon de Chelly: the Story behind the Scenery*. KC Publications, Las Vegas, Nevada.

Canyonlands National Park

Established: 1964
Size: 1367 square kilometers (527 square miles)
Elevation: 1125 to 2006 meters (3680 to 6560 feet)
Address: Moab, Utah 84532

STAR FEATURES

• Well-exposed "layer cake" geology with vivid erosion-carved tiers of Mesozoic sedimentary rocks.

• A dramatic landscape of deep canyons separating mesas, buttes, and pinnacles shaped by water, wind, and gravity.

• Two great rivers, masters of canyon country erosion.

• Anticlines, grabens, and other geologic structures created by slow-flowing subterranean salt.

• An unusual and interesting dome resulting from meteorite impact.

• Visitor center, guided tours, trail and roadside exhibits, guide leaflets, and evening programs.

See color pages for additional photographs.

SETTING THE STAGE

Where two mighty rivers meet, they have shaped a lonesome, dramatic land of barren canyons, cliffs, buttes, and mesas, a strangely beautiful, awesome valley where erosion rules. This is harsh country, dissected by tortuous canyons that are obstacles to travel and exploration. Yet it is fascinating country, because here the geology is written boldly and clearly on the denuded landscape.

Sedimentary rocks in this national park, flat-lying or very gently arching across the anticline of the broad Monument Uplift, have eroded into layered scenery where alternating bands of resistant sandstone and less resistant siltstone form a ledge-slope-ledge topography. Above particularly resistant cliff-formers, slopes flatten into nearly horizontal benches, natural landings between the plateau surface and the rivers.

Many cliff-forming sandstones show the broad, sloping cross-bedding and fine, rounded and frosted grains that characterize dune sand. Silt-stone and mudstone slope-formers, deposited in ancient deltas or on wide river floodplains, still bear ripple marks and raindrop impressions formed before they hardened into rock. Fossils are extremely rare except in a few thin limestone layers in the lowest part of the rock sequence.

Almost all of these rocks are pink, brick red, or salmon colored, tinted with varying amounts of hematite. Some rock surfaces are coated with dark brown or blue-black desert varnish.

In regions of flat-lying, uniform strata, streams normally develop dendritic drainage patterns, branching like trees. Here this pattern is carried to extremes. Tributary stream courses divide and redivide until they are the fine rills that decorate the highest slopes of the canyon walls. Wind, gravity, and short-lived torrents fed by sudden

Permian rocks of the Needles area erode along vertical joints. Many layers thicken, thin, or disappear over short horizontal distances. Ray Strauss photo.

The Island in the Sky looks down on a broad bench of White Rim Sandstone, above the scalloped margins of an entrenched meander of the Green River. Ray Strauss photo.

storms wash and wear away siltstone and mudstone, and undermine sandstone cliffs so that great rock slabs fall away along vertical joints. Century by century, canyons widen and plateaus reduce to mesas, mesas to buttes, buttes to pinnacles, until nothing is left but piles of rocky rubble. Eventually the rubble breaks up too, and is blown away or carried by storm-born streams to the two great rivers that flow through the park.

The Colorado and Green Rivers, with headwaters in the Rocky Mountains of Colorado and Wyoming, regulate the depth of side canyons. No tributary can erode below the level where it joins these main streams. Placid as they enter the park, the two rivers join forces to rampage through Cataract Canyon, one of the wildest stretches of water in any national park. Furthermore, the two rivers follow twisting courses that reveal oddities of their earlier history, when they meandered over a wide plain covered with Tertiary sediments.

Along the canyons of the Green and Colorado, strata bow up slightly toward the river, partly because of the gentle, almost imperceptible arch of the Monument Uplift, and partly in response to the upward push of thick layers of salt some distance below the surface. Salt becomes plastic under pressure, flowing extremely slowly, moving from regions where pressure is great to regions of lower pressure. Pressure on the salt is mostly due to overburden, the weight of 3000 meters (10,000 feet) or more of Paleozoic and Mesozoic strata, plus a great thickness of Tertiary sediments now eroded away completely.

Where the Green and Colorado Rivers and their larger tributaries removed some of this overburden, the pressure on the salt lessened. And so through thousands and even millions of years, the salt flowed toward the rivers, bowing up the strata so gradually that the rivers never fluctuated from their courses. (An alternate explanation of

In the Needles District, where sets of parallel joints intersect at almost right angles, the Cedar Mesa Sandstone erodes into a forest of fantastic pinnacles. Ray Strauss photo.

the salt anticlines along the rivers places them directly over large faults in Precambrian rocks, as at Arches National Park. The salt would in this case flow toward the fault ridges, forming anticlines above them. The rivers would later establish courses along valleys caused by solution of the salt and collapse of overlying rocks.)

GEOLOGIC HISTORY

Paleozoic Era. For much of Paleozoic time this area lay beneath a shallow sea, where successive layers of marine limestone, sandstone, and shale were deposited. None of the older marine strata appear in the park, but thick beds of Pennsylvanian salt influence the present park picture, so let's start our discussion with their deposition.

In Pennsylvanian time, this area was inundated by a shallow sea. A high fault-block island rose in what is now western Colorado, restricting circulation of the sea. At the same time, a submerged platform on the sea floor west and southwest of Canyonlands also restricted circulation. Between island and platform, in an often landlocked embayment, sea water evaporated. As the water became briny and finally dried completely, common salt, potash, anhydrite, and gypsum—all water-

soluble—were deposited. Twenty-nine distinct layers of these minerals have been identified in western Colorado and eastern Utah in oil well cores; their total thickness is nearly 1000 meters (3000 feet). Between the salt layers are thin black and brown shales that indicate stagnant conditions, just what one might expect in such a restricted basin.

Permian sedimentary rocks above the salt record filling of the basin, as well as the presence of nearby highlands, sources for their sand and conglomerate pebbles. In Permian time the area was occupied alternately by shallow sea and by beaches and dunes in which cross-bedded sandstone was deposited.

Mesozoic Era. In Triassic time rivers and streams brought mud, silt, and sand from the uplift in Colorado, and deposited them in broad, thin sheets on their floodplains and deltas. The geographic scene may have been similar to that near the mouth of the present Mississippi or the Nile, but with nearby mountains and without evidence of man or other mammals.

New mountains rose in Jurassic time in California and central Utah, serving as added sources of sediment. They intercepted moisture-bearing winds from the west, so that the Canyonlands region, as well as other parts of the Southwest, became deserts. A vast sea of dunes developed here, rivaling (and resembling) the deserts of North Africa and the Arabian Peninsula today. Many of the dunes were separated by flat interdune hollows where silt and clay accumulated. The dune sands exist today as cross-bedded, cliff-forming sandstones, some of them with thin horizontal bands of finer, redder siltstone representing interdune deposits. Dune sandstone is recognized by its light color, very large-scale cross-bedding, abrupt boundaries, and even-sized rounded and frosted sand grains. In the southwestern United States, an incredible amount of dune sand accumulated in Mesozoic time.

The rocks you see in Canyonlands record both Triassic and Jurassic dunes. Not until Cretaceous time did seas once again invade the area. Thick marine shale and sandstone layers deposited in those seas have now been stripped away by erosion, but they are known from such nearby areas as Mesa Verde National Park.

Cenozoic Era. After Cretaceous time the region rose above sea level again. Although there is no record within the park of Tertiary deposits, we know from adjacent regions that thousands of meters of silt, clay, and fine sand stripped from the newborn Rocky Mountains were deposited here in early Tertiary time. Many of these rocks contain glassy shards of volcanic ash, reflecting volcanic activity in the Yellowstone region and elsewhere. The Colorado and Green Rivers established courses across the broad plains below the mountains, coming together in what is now Canyonlands National Park.

In Miocene and Pliocene time the whole Rocky Mountain region, including eastern Utah, bowed upward, rising as much as 2000 meters (6000 feet) above its former position. Rivers and streams, their courses steepened, scoured ever more deeply, washing away the fine, unconsolidated silt and sand and volcanic ash deposited in early Tertiary time. In these soft sediments, the Colorado and Green Rivers deepened their meandering channels. Soon imprisoned in steep-walled gorges, they nevertheless retained their twisting, winding courses. And with the help of thousands of small tributaries, most of them flowing only after heavy rains, these great rivers carved Canyonlands as we see it today.

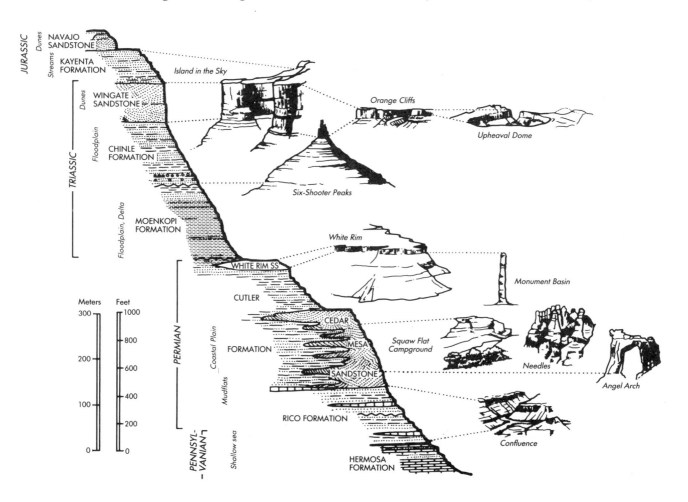

Druid Arch formed in a thin fin of Cedar Mesa Sandstone. Ray Strauss photo.

BEHIND THE SCENES

Cataract Canyon. Viewed from the rim of the canyon above the Confluence, the Colorado and Green Rivers flow smoothly, with gradients of about 0.2 meter per kilometer (1 foot per mile), gradually blending their waters. But several kilometers downstream the combined rivers—now called the Colorado—surge into Cataract Canyon, named by John Wesley Powell in 1869 as he led the first Colorado River float trip. There the Colorado plunges again and again through churning rapids caused by rock debris washed in from tributary canyons, by massive talus cones below vertical canyon walls, and by huge fallen boulders that constrict the stream. More than half the rapids encountered by Powell are now drowned beneath the waters of Lake Powell.

Along parts of Cataract Canyon, rock strata dip away from the river because of the rise of buried salt. Rocks near the river are the oldest in the park: Permian marine limestone and, below it, fossil-bearing Pennsylvanian marine limestone, sandstone, and shale. Pennsylvanian salt beds are less than 1000 feet below the surface. Gypsum, less soluble than salt but like it becoming plastic with pressure, breaks through to the surface in the depths of Cataract Canyon near the mouths of Red Lake and Gypsum Canyons.

Colorado and Green Rivers. Above their confluence the Green and Colorado Rivers are docile and easygoing, essentially free from rapids. They swing back and forth in meanders inherited from early Miocene time, when they looped across a nearly horizontal plain. The rivers have shortcut some of their entrenched meanders, leaving abandoned loops high above present river level. The most prominent of these is outside the park: Jackson Hole, across the Colorado River from Potash. Anderson Bottom, on the Green River, is another.

The Confluence. The junction of the Green and Colorado Rivers is often made interesting by differences in the amount of sediment they carry. Sometimes the Colorado is muddier than the Green; sometimes it is clearer. For some distance after they join, the clear and muddy waters flow side by side, unmixed. However, irregularities in the channel cause turbulence and eventual mixing.

In times of low water, sandbars and beaches develop here. Salt cedars (tamarisk) growing on the sand have stabilized many of these bars and beaches. Salt cedar is a newcomer, a native of the Near East, introduced into Mexico. It has now invaded most Southwestern rivers, particularly flood-free ones controlled by dams, thriving wherever it is well fed by river or spring waters. Native plants can't compete with it, and its introduction reduces natural forage for wildlife. It is a prodigious water user, able to dry up natural springs.

The Grabens. Accessible only on foot or by four-wheel-drive vehicle, the Grabens are slender, flat-floored chasms as much as 100 meters (300 feet) deep. In the Needles Fault Zone, the Grabens occur where parallel vertical faults separate slim slices of rock. There are quite a number of these linear valleys here; four are traversed by four-wheel-drive trails: Devils Canyon, Devils Lane, Cyclone Canyon, and the unnamed graben leading to Beef Basin.

The origin of the Grabens lies in movement of subterranean salt out of this area and into the salt-cored anticline below the nearby Colorado River. Grabens are due to tension, a stretching and breaking of the layered rock as the salt anticline formed. The Grabens are geologically young; many still have smooth, essentially vertical fault-scarp walls almost untouched by erosion. Cyclone Canyon and several smaller grabens have not yet developed through drainage.

Grand View Point. From this high, readily accessible vantage point, with magnificent views out over almost all of Canyonlands National Park, you can easily identify the rock layers by their expression as cliffs, ledges, slopes, and wide benches. White Kayenta Sandstone surfaces the point; brick red mudstone of the Rico Formation appears near the rivers. Notice particularly the tall cliffs of Wingate Sandstone, dark with streaks of lichens and desert varnish, and the conspicuous wide bench of White Rim Sandstone at the top of the Cutler Formation. The White Rim Sandstone thickens westward, forming a higher and higher cliff below this bench. The tiny track winding along the bench is the White Rim Road, suitable only for four-wheel-drive vehicles. To the southwest the Green River negotiates a hairpin turn in one of its entrenched meanders.

In Monument Basin, southeast of the overlook and about 3 kilometers (2 miles) away, below the White Rim bench, a tall finger of Cutler Formation indicates the scale of the surrounding country. You can see about two-thirds of this rock totem pole, which is 94 meters (305 feet) tall and less than 12 meters (38 feet) thick.

Three mountain ranges break the skyline: the La Sals to the northeast, the Abajo Mountains to the southeast, and the Henry Mountains to the southwest. The three ranges formed as igneous rock welled upward in mid-Tertiary time, doming sedimentary layers, but not, as far as we know, breaking through to the surface. On the horizon just west of the Abajo Mountains, a broad arch of sedimentary rocks extends south toward Arizona.

Between here and the Abajo Mountains you can also distinguish the finely divided red spires of the Needles District of the national park.

Island in the Sky. Riding the wedge between the Green River on the west and the Colorado on the east, Island in the Sky rises 600 meters (2000 feet) above the two rivers, offering excellent views of the whole Canyonlands area, an incredible maze of benches, canyons, buttes, pinnacles, and brilliant-hued rocks in an all-pervading layer-cake pattern.

Island in the Sky is a Y-shaped mesa standing almost free of the country to the north. You enter the Y along its eastern arm, crossing the slender "Neck" that is its only connection with the "mainland." Ultimately, as cliffs on either side are undermined by erosion, the Neck will fall, isolating the Island completely.

Near its edges the Island is surfaced with ledgy layers of Kayenta Sandstone; in its highest portions there are rounded knolls and knobs of cross-bedded, white Navajo Sandstone, a thick dune sandstone that covers much of Utah, northeast Arizona, and western Colorado. High cliffs that drop off vertically all around the Island are the Wingate Formation. Lower layers are similar to those seen from Grand View Point.

The Maze, Land of Standing Rocks and Elaterite Basin. The part of the national park west of the rivers can be reached only by four-wheel-drive vehicles. The Land of Standing Rocks is part of a broad bench on the Cedar Mesa Sandstone, the lowest part of the Cutler Formation. Stream patterns are dendritic in most of this area, but controlled by joints in such places as the Fins. Several

With its seemingly structureless central mass and its concentric rings of Triassic and Jurassic rock, Upheaval Dome took shape when a meteorite struck the Earth's surface, probably in Cretaceous time. National Park Service photo.

rock units, including the White Rim Sandstone and parts of the Moenkopi Formation, thin or thicken between layers of siltstone. Such changes are common; they show up particularly well here because of contrasts in rock color. They represent horizontal changes in deposition, where the kind of sediment changes from place to place, as in beaches and muddy lagoons, dunes and interdunes, or sandy channels in silty floodplains. Watch for these features and for faults, which show up as offsets of the horizontal rock layers.

Elaterite Basin, mostly outside the national park, and Elaterite Butte, within the park, are named for a dark brown tarlike mineral that seeps from parts of the White Rim Sandstone. Elaterite originates from organic matter, as does its relative, oil, and may travel some distance from its source to lodge in a porous reservoir rock.

The Needles District. The broad, level surfaces of the Needles District are part of a wide bench between the rim of the canyon country and the inner chasm of the Colorado River. The rocks that surface most of the bench are the Cedar Mesa

Sandstone, a striped red and white unit in the lower part of the Cutler Formation. The striped appearance of this rock results from interlayering of the red eastern part of the Cutler Formation with the white sandstone that characterizes the same formation farther west.

The White Rim Sandstone, which makes a prominent bench in the northern part of Canyonlands, is not present here; bright vermilion siltstone and sandstone occupy its position at the top of the Cutler Formation. These rocks appear in the lowest, orangest slopes of North and South Sixshooter Peaks, and at corresponding levels in the cliff facade to the east. Higher parts of these slopes are the Moenkopi and Chinle Formations. The cliffs themselves, darkened with desert varnish, are Wingate Sandstone.

The Needles District owes its carved turrets and occasional arches to weathering and erosion of Cedar Mesa Sandstone along two intersecting sets of joints and faults, one paralleling the Grabens, the other at almost right angles. In the enchanted world of the Needles themselves, joints are particularly closely and regularly spaced. Dune sandstone such as that of the Cedar Mesa Sandstone shows well-sorted and evenly rounded grains; its smooth texture lets it break away, where it is undermined, in caves and natural arches. All the arches in this part of the park are in this sandstone. In places, faults offset the red and white rock layers.

The Potholes. Where the undulating surface of the Cedar Mesa Sandstone collects rainwater, shallow pools nourish tiny plants and animals born of spores, eggs, and seeds that can survive long droughts between rains. Small shrubs and grasses grow nearby, their roots searching rock joints for precious moisture. Metabolic acids secreted by these organisms add to the slight natural acidity of rainwater, so that the water in pools and crevices dissolves the calcium carbonate cement that holds together the sand grains of the sandstone. Little by little, grain by grain, helped by wind and frost, slight depressions become deep potholes, more and more able to hold rainwater and so to perpetuate the cycle.

Salt Canyon and Horse Canyon. Four-wheel-drive trails threading these canyons, and foot trails branching from them, lead to Paul Bunyans Potty, Gothic Arch, Castle Arch, and Angel Arch, as well as to prehistoric rock art and small ruins. (For a discussion on the origin of arches, see Arches National Park.)

Horse Canyon is relatively straight and shallow, sand floored, and walled only with Cedar Mesa Sandstone. Its drainage area is small, rarely providing enough water to clear the canyon down to bedrock.

Salt Canyon, on the other hand, is winding and rock floored. Its drainage area is larger, and the canyon as a result cuts through the Cedar Mesa Sandstone and into the underlying Rico Formation, following a twisting course along joints related to the Needles fault zone.

Shafer Trail. This four-wheel-drive road, built for access to uranium mines, zigzags dizzily down the vertical bastion of Island in the Sky, dropping through almost all the rock units exposed in the park. Ledgy cliffs of Kayenta Sandstone, and below them a precipice of Wingate Sandstone, border the road's initial descent. Below the cliffs, the route provides a good look at the Moenkopi and Chinle Formations just above the White Rim. Some of the mudstones of these formations bear ripple marks and raindrop impressions formed 200 million years ago. In canyons and gullies below the White Rim are red and purple sandstone and conglomerate of the Cutler and Rico Formations. A few thin limestone layers in the Rico contain fossil shellfish of very late Pennsylvanian and early Permian age.

Upheaval Dome. Long thought to be due to flowing salt pushing up toward the surface, this startling circular structure—a mound of Triassic sandstone and mudstone surrounded by concentric rings of much-faulted younger rocks—is now believed to be the result of a meteorite impact. At the time of impact the rocks you see here were deeply buried; at least 1000 meters (3000 feet) of overlying rocks have been removed by erosion since the dome formed.

The impact and subsequent explosion are thought to have created a short-lived underground cavity 6 kilometers (4 miles) or more in diameter. Such a cavity would immediately collapse, with its walls, already broken by the shock wave of the impact, sliding inward along curving faults, meeting and pushing upward at its center. The meteor crater formed at the surface, long since eroded away, was probably 8 or 9 kilometers (5 or 6 miles) in diameter.

Study of the rocks distorted by the impact, and projection of the strata that once lay above them, suggest that the impact occurred in very late Cretaceous time or very early Tertiary time, possibly as part of the million-year-long meteor shower that caused widespread extinction of plant and animal species at the Cretaceous-Tertiary boundary.

White Rim Trail. This four-wheel-drive track follows a tortuous course from Potash on the Colorado River south past the base of the Shafer Trail and Grand View Point, and then north to Horsethief Trail, outside the park on the Green River. For almost all of this distance the track rides the upper surface of the White Rim Sandstone at the top of the Cutler Formation. Strata visible along the route are the same as those described for Grand View Point.

OTHER READING

Lohman, S.W., 1974. *The Geologic Story of Canyonlands National Park*. U.S. Geological Survey Bulletin 1327.

Dune sandstone of the Navajo Formation forms the mesa surface near Elephant Butte and the Windows Section, where arches are carved in Entrada Sandstone. Distant La Sal Mountains rose as hot magma pushed upward in clustered laccoliths.

Sweeping cross-bedding of the Navajo Sandstone developed on Jurassic sand dunes.

Undermined by erosion of the soft, irregularly bedded Dewey Bridge Member, Double Arch formed in the Slick Rock part of the Entrada Formation.

Bryce Canyon National Park

Hikers view aspects of Bryce not seen from the rim. Note slanting scour marks of an earlier soil level at the base of the wall at the left.

Below, shaped by wind and water, the Natural Bridge reflects luminous high-altitude sunshine.

The dramatic escarpment at Cedar Breaks is caused by headward erosion of streams on the western escarpment of Utah's High Plateaus. (Zion Natural History Association photo.)

Where the Green and Colorado Rivers join, their separate waters flow briefly side by side. In this photo the Green is brown, the Colorado is green. Color depends on upstream storms that wash mud into one or both rivers.

A wide, flat-floored vale near its southern end, the Devils Lane graben narrows northward to less than 3 meters (9 feet).

Erosion widens closely spaced joints in the Cutler Formation, creating the red-and-white striped pinnacles of the Needles District.

The upward-pointing barrel of Sixshooter Peak formed as Wingate Sandstone broke away along vertical joints. Sandstone and siltstone of the Chinle Formation color symmetrical slopes below.

Capitol Reef National Park

Blue-black desert varnish, glinting in the sun, surfaces a canyon wall.

Below, *Hickman Bridge is one of several natural bridges in the national park. Bridges span stream courses, arches do not.*

Navajo Sandstone, Triassic-Jurassic in age, erodes into the white domes that give Capitol Reef its name.

Seen from Desert View, bright red Precambrian sedimentary rocks are beveled and covered by horizontal Cambrian strata.

From the ancient gneiss of the Inner Gorge to the white limestone of its rim, Grand Canyon unfolds 2 billion years of Earth history.

Once the playthings of the untamed river, boulders such as these are no longer swept downstream.

This 500-foot precipice gives its name to the Redwall Limestone. Really gray, the limestone is painted with an iron oxide wash from red slopes above.

Grand Canyon National Park

Wide, placid waters funnel toward the narrow Inner Gorge, where hard Precambrian rocks confine the river and increase its speed and turbulence. The ancient rocks are beveled by the Great Unconformity.

For much of the year, the Little Colorado flows turquoise waters from springs some distance above its mouth. Magnesium compounds add chalkiness that brings out the natural blueness of the water.

Pink pegmatite veins lace the Vishnu Schist. Their color comes from long-term bombardment by radioactive minerals inherent in the rock.

Sections of petrified trunks, broken along joints, reveal their cellular structure and the colorful agate and jasper with which they are impregnated.

Blue Mesa's brown slopes are paved with pebbles fallen from the conglomerate at the right. Many of the pebbles came from Triassic mountains in central Arizona. Blue, gray, and white banding marks volcanic ash layers.

Cloud shadows accent the Painted Desert's slopes, colored with iron and manganese oxides. White deposits along dry watercourses are lime (calcium carbonate) and gypsum.

Zion National Park

Below, *stream erosion brings out details of cross-bedding of the Navajo Sandstone (Lucy Chronic photo). Right, potholes in Echo Canyon were ground by rock millstones whirled by a rushing stream.*

Left, *in dry weather, Zion Canyon's sheer-walled Narrows invite hikers. They are out of bounds, though, during the summer thunderstorm season (Felicie Williams photo). Above, Zion Canyon, seen here from Observation Point, exposes the great thickness of Navajo Sandstone.*

Capitol Reef National Park

Established: 1937 as a national monument, 1971 as a national park
Size: 978 square kilometers (378 square miles)
Elevation: 1676 meters (5500 feet) at visitor center
Address: Torrey, Utah 84775

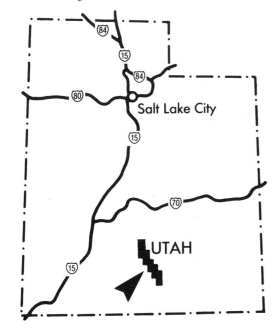

STAR FEATURES

• A dramatic monocline in otherwise nearly flat-lying sedimentary rocks.

• Narrow, high-walled canyons, rock arches and natural bridges, and "waterpockets" eroded in dune-formed sandstone.

• A well-exposed sequence of color-splashed Mesozoic sedimentary rocks carrying records of floodplain, delta, tidal flat, and desert environments.

• Indirect evidence of Pleistocene glaciation.

• Visitor center, museum, scenic drives and trails (some with guide leaflets), guided hikes, and evening programs. A relief map in the visitor center presents an excellent overview of the park and its geology.

See color pages for additional photographs.

SETTING THE STAGE

In this national park, the major scenic feature is a single large monocline known as Waterpocket Fold. Involving layered sedimentary rocks of Triassic, Jurassic, and Cretaceous age, steeply dipping toward the east, this fold measures nearly 160 kilometers (100 miles) in length. It is only a fraction that wide: less than 5 kilometers (3 miles) through most of its length. Along it, nearly horizontal rock layers suddenly plunge eastward like a breaking wave, and then flatten out again at a lower level. Because each rock type responds differently to erosion, a parallel array of ridges and valleys marks the position of the fold. The highest ridge, in the most massive and resistant rock layers, is known as Capitol Reef.

The name comes from several rounded, bare-rock summits that resemble capitol domes, coupled with an old word for a rocky barrier (at sea or on land). And a barrier it is, with only four canyons (all of them subject to flash floods) allowing passage for water, animals, and man. The name of Waterpocket Fold stems from natural rain-filled pools, many of them pocket-shaped, in the massive sandstone at the heart of Capitol Reef.

Waterpocket Fold extends from Thousand Lake Mountain just north of the national park, south and southeastward to Lake Powell. Along its west side the fold is marked by a high, angular cliff of Wingate Sandstone, red tinted and daubed and streaked with lichens and desert varnish. The cliff rises sharply above slopes of soft green and purple shale and volcanic ash of the Chinle Formation, which in turn lies on dark red sandstone and siltstone of the Moenkopi Formation, the surface rock in western parts of the park. Erosion of the Moenkopi and Chinle Formations undermines the massive cliffs, so that from time to time giant slabs, no longer supported at their base, break off along vertical joints and crash to the slope below.

Above and east of the cliff-forming Wingate Sandstone, separated from it by thinner layers of the Kayenta Formation, is another thick unit, the Navajo Sandstone. Not as well compacted or cemented as the Wingate, the Navajo Sandstone weathers into rounded cliffs and uplands—the barren "slickrock" of the summit of Waterpocket Fold. East of the summit, the denuded upper surface of this formation is fully exposed as the east flank of Capitol Reef. Thus the reef itself is asymmetrical: angular cliffs on the west, rounded bare-rock slopes on the east.

Both the Wingate and the Navajo Sandstone are marked with the long, sweeping cross-bedding and have the fine, rounded and frosted grains of dune-formed sandstone. Wind-formed ripple marks decorate many rock surfaces. (Unlike ripple marks formed by water, which are always at right angles to the stream flow, dune ripples may run up and down dune surfaces.)

There are no in-place lava flows within the park, but near the Fremont River boulders of dark gray lava veneer flat-topped hills. Well rounded by stream action, these boulders were obviously carried for some distance. Their presence adds an interesting note to the geologic story of this area. In Cathedral Valley, in the northern part of the park, dikes and volcanic necks mark sites of Tertiary volcanism.

Though only four canyons penetrate the reef's entire thickness, the reef itself is a labyrinth of canyons, some sheer-walled and sandy-floored. Important factors in their development are day-night temperature contrasts and the cloudburst climate of this part of the Southwest. Temperature contrasts that cause freezing and thawing of small amounts of moisture, either in joints or between sand grains of porous rock, are insignificant in the short term. But repeated again and again they eventually loosen sand grains and break off rock flakes, initiating breakdown of solid rock. Such frost wedging is especially effective in shady rock clefts and narrow defiles, where moisture is retained for longer periods.

Frost may initiate development of waterpockets, too, enlarging slight depressions created by wind erosion. As the depressions deepen, as they hold more water for longer periods of time, spores and seeds of small plants and larvae of tiny animals are brought to them by wind or by birds, lizards, and other animals coming to the pools to drink. Developing in the pools, the little plants and animals, many of them microscopic, secrete acid metabolic products that dissolve the calcium carbonate cement of surrounding sandstone. Then, when the pools dry up, wind blows away the loose grains, deepening the hollows.

Small rock depressions may string together, particularly along joints, guiding the flow of rainwater, inaugurating development of clefts, crevasses, and eventually, deep, narrow canyons. Water from sudden downpours churns through these passageways, scouring and smoothing their floors and walls. The swirling streams may also grind out large potholes, or carve arches and natural bridges. (For more on arch formation, see Arches National Park.)

GEOLOGIC HISTORY

Sedimentary rocks in this park range from Permian sandstone and marine limestone in several deep canyons along the west edge of the park, to Cretaceous shale and sandstone in parallel valleys and ridges east of Capitol Reef. Representing a time span of about 200 million years, these rocks were deposited in a long succession of changing environments.

Paleozoic Era. The story of this park begins 260 million years ago on a Permian desert, with sand dunes that have become the Cutler Sandstone, now visible in the bottoms of the deeply entrenched canyons of Sulphur and Fremont Creeks in the western part of the park. Above the Cutler Sandstone are light gray ledges of Kaibab Limestone, the rimrock of the Grand Canyon, here sandy and impure because it was deposited close to the shore of the Permian sea, the last Paleozoic sea to transgress the continent.

Mesozoic Era. In Triassic time a large marine embayment covered much of southern Utah, opening westward to the sea. In the shallow bay, successive layers of sand and mud accumulated, to become the dark red sandstone, siltstone, and shale of the Moenkopi Formation. Many characteristics of these rocks suggest origin on tidal mudflats: well-preserved ripple marks, mud cracks, salt crystal impressions, gypsum (from evaporation of sea water), and alternation of mudstone, siltstone, and sandstone. A few limestone layers, however, contain fossils of marine shellfish. The formation is nearly 300 meters (1000 feet) thick just west of the visitor center; it thins southeastward and finally disappears altogether. Sand, silt, and mud of the Moenkopi strata are thought to have come from highlands in western Colorado.

Above the Moenkopi Formation, the green and purple Chinle Formation was also partly deposited by water—probably fresh water. It contains a high proportion of volcanic ash decomposed into bentonite, a combination of clay minerals that swell when they are wet, shrink when they are dry. The puffy, crumbly surface produced by these changes is easily blown away by wind or washed away by rain.

The bentonite tells us that there must have been volcanoes somewhere in this area in Triassic time. And eruptions must have been voluminous, for the bentonite-rich part of the Chinle Formation spreads across southern Utah and northern Arizona. It is the rock of Arizona's Painted Desert and Petrified Forest National Park. The Chinle Formation also includes several layers of hard, ledge-forming sandstone and conglomerate derived, it is thought, from highlands in central Arizona.

As volcanism died down in late Triassic time, a sandy desert developed here, one that rivals the modern Sahara and Arabian Deserts. Swept by winds, dunes built up across the land, later to become the Wingate Sandstone. Above the Wingate cliffs are thinner river-deposited sandstones sep-

Colorful Triassic rocks, from the Moenkopi Formation (bottom) to the Wingate Sandstone (top), form the bold western face of Capitol Reef. Ray Strauss photo.

arated by horizontal bands of fine siltstone: the Kayenta Formation.

In Jurassic time, roughly 190 million years ago, dunes built another massive deposit—the Navajo Sandstone, the rock that rides the crest of Capitol Reef and shelters the many waterpockets of this park.

Rocks younger than the Navajo Sandstone appear in ridges and valleys along the east side of Waterpocket Fold. Limestone, sandstone, mudstone, and gypsum of the Carmel Formation suggest several incursions of the sea. The Entrada Sandstone, looking rather like the Moenkopi on the west side of the fold, and the harder gray Curtis Formation above it were formed on floodplains and tidal flats; above them the Summerville Formation represents a lake environment. The youngest Jurassic unit is the Morrison Formation, its bentonite-bearing shales as colorful as those of the Chinle Formation west of Capitol Reef. Fossil dinosaur bones have been found in the Morrison Formation here and elsewhere.

In Cretaceous time the land sank below sea level again. A thin layer of near-shore sandstone—the Dakota Sandstone—underlies a 1000-meter (3000-foot) sequence of gray marine shales —the Mancos Shale. The Dakota Sandstone in places contains rounded pebbles that indicate strong wave or current action; elsewhere it contains fossil oyster shells. East of the park boundary, these rock layers flatten out perceptibly, and the gray Mancos Shale, weak and easily eroded, weathers into badlands. Farther east, forming a mesa, are beach and near-shore deposits of the Mesaverde Group. These Cretaceous rocks bring the Mesozoic Era to a close. To the west the Wasatch Mountains were rising, and to the east the Rockies. Both would bring many changes to this area.

Cenozoic Era. As the Colorado Plateau pushed upward, probably in fits and starts, its individual blocks, some raised higher than others, remained essentially horizontal. On the highest plateaus, erosion bit most deeply, stripping away many of the sedimentary strata. Hard limestone layers and well-cemented sandstone remained as plateau surfaces or, along folds and faults, stood as lines of cliffs.

There are no Tertiary sedimentary rocks in this park. Some undoubtedly once lay across this area: soft lake siltstone and limestone similar to that at Bryce Canyon National Park. But like the thick Cretaceous layers, they long ago eroded off Capitol Reef.

Black lava boulders on Johnson Mesa, near the park headquarters, tell of Tertiary volcanism and Pleistocene glaciation west and north of the park. The boulders, some of them a meter or more in diameter, are composed of 20-million-year-old lava quarried about 25,000 years ago by Ice Age glaciers on high plateaus to the west. Streams gushing from melting glaciers tumbled the boulders along for many kilometers, until blocked by the partial barrier of Capitol Reef. Some were conveyed through the Fremont River's narrow canyon and dropped on the east side of the reef.

BEHIND THE SCENES

Capitol Dome. High on the crest of Capitol Reef just north of the Fremont River, Capitol Dome and several other dome-shaped summits are shaped in pale gray or yellowish Navajo Sandstone. Sweeping cross-bedding of the dune sandstone marks the domes and rocky uplands surrounding them. Narrow creases that run down the sides of the domes are channels eroded by rain.

Capitol Gorge. At the western, upstream end of this narrow gorge, Wingate Sandstone cliffs, tapestried with lichens, mineral stains, and glossy, blue-black desert varnish, rise from the streambed. The vertical walls also display large-scale, dune-style cross-bedding.

Pits and hollows of honeycomb weathering are common here. Starting with tiny depressions caused by solution of the calcium carbonate that holds the sandstone together, these pits are partly the work of wind, which helps to deepen them by whirling the loosened sand grains and eventually removing them. Nearby, deep, smooth scallops in the rock wall are remains of potholes created by swirling stream water that spun rock against rock. The stream later sliced the holes in two as the canyon deepened.

Farther into the gorge, the Pioneer Register records early travelers who engraved their names in the desert varnish that coats the rock. (Some of these signatures are out of reach even from a wagon seat, and are thought to have been carved by pranksters lowered over the cliff by ropes.) Storm waters funneling through the narrow canyon frequently change the level of its sandy floor. Between 1880 and 1962 a road through Capitol Gorge was the only highway between Torrey and Hanksville; road repairs were necessary after every storm.

Still farther east, the Wingate Sandstone plunges beneath the surface, and the Kayenta Formation and Navajo Sandstone wall the canyon. Because these units are a little more easily eroded

Eons of erosion by wind, water, and frost have shaped the deep gorges and high barren summits of Waterpocket Fold. Ledges and slopes are in the Kayenta Formation, with Navajo Sandstone above and Wingate Sandstone below. Ray Strauss photo.

they stand in less formidable cliffs. In them several waterpockets, known as the Tanks, usually contain standing water from the last storm. They lie along rock joints that were natural pathways for water sheeting off rock surfaces above. As channelways deepened, water collected behind cross-bedding ridges in the sandstone. Small pools became larger as acids secreted by tiny organisms dissolved the cementing material that held the sand grains together.

A trail from Capitol Gorge climbs to the base of the Golden Throne, one of the massive Navajo Sandstone monuments that crown Capitol Reef.

Cassidy Arch. The trail to Cassidy Arch climbs steeply from Grand Wash through the Wingate Sandstone, then levels off among rounded erosional hummocks of the Navajo Sandstone. Swirling cross-bedding marks barren uplands; overhangs and blackened hollows are incipient arches and waterpockets. Cassidy Arch itself, a graceful rainbow, is one of these features.

Cathedral Valley. At the northern end of the national park, this valley is east of Waterpocket Fold. Here, in a maze of mesas and cuestas, 600-meter (2000-foot) cliffs overlook a broad, eroded surface dotted with cathedral-like buttes. The road into the valley passes through the Bentonite Hills in the Morrison Formation, part of Utah's Painted Desert, to reach overlooks above Cathedral Valley.

Shaped by wind and water, the great amphitheater is ornamented with unusual erosional forms carved in rocks younger than those of Capitol Reef—the Entrada Sandstone, of Jurassic and Cretaceous age. Dikes, sills, and volcanic necks jut above the surface. A gypsum sink results from solution of underground gypsum.

Chimney Rock. Shaped in red mudstone and siltstone of the Moenkopi Formation, this high tower stands out from a promontory of similar rock. Both are capped with hard, sandy, pebble-filled Shinarump Conglomerate. As you can see from slopes on the other side of the road, the Moenkopi is not normally a cliff-former. But here where patches of Shinarump Conglomerate protect it and rapid erosion undermines it, it stands in almost vertical walls.

A fault separates these walls from the mesa to the north, where the Shinarump Conglomerate is at the base of the mesa cliff. Displacement along the fault is about 50 meters (165 feet). Petrified wood is common in the Chinle Formation here—not the prized agatized wood of Petrified Forest

National Park in Arizona, but a duller variety impregnated with less colorful varieties of silica.

Ripple marks, mud cracks, and raindrop pits mark rock slabs of the Moenkopi Formation. The dark red color of these rocks is due to oxidation of iron minerals after the rock was deposited. White or pale pink veins are gypsum.

Cohab Canyon. The trail to Cohab Canyon climbs steeply through soft purple and green shale of the Chinle Formation, rich in clay minerals derived from volcanic ash. It then skirts towering cliffs of Wingate Sandstone, marked with black, yellow-green, and white mineral stains. Pronounced vertical fractures govern the shape of the cliffs. Some rock slabs have broken away in arcs, in shell-like (conchoidal) fracture patterns.

Cohab Canyon's narrow defile has eroded headward from east to west, and drains into the Fremont River. Along the trail watch for sedimentary and erosional details: ripple marks on flat rock faces, potholes, honeycomb weathering, and liesegang rings—decorative concentric brown arcs and bands caused by precipitation of iron oxide from groundwater.

Fremont River. Heading in Fish Lake in the Fishlake Mountains northwest of Waterpocket Fold, the Fremont River transects the entire sequence of east-tilted Mesozoic rocks, from Moenkopi Formation redbeds near the visitor center to gray Mancos Shale east of the fold. Utah Highway 24 follows the river's route. Rock layers—the Wingate Sandstone, Kayenta Formation, and Navajo and Entrada Sandstones—are tilted most steeply at the west side of the fold, and progressively more gently to the east. Farther east are the Bentonite Hills of the Morrison Formation —Utah's Painted Desert—and a sloping cuesta capped with tan Dakota Sandstone. Drab gray hills of Mancos Shale lead to mesas capped with Mesaverde Sandstone.

The waterfall near highway mile 86-87 is man-made, formed when highway construction blocked a gooseneck loop of the river. Black lava boulders, brought here by streams draining Ice Age mountain glaciers, occur along the river for some distance east of the narrow part of the canyon.

Goosenecks of Sulphur Creek. Viewpoints at the Goosenecks look down on the oldest rocks in the park: buff-colored Cutler Sandstone, a Permian dune deposit, and above it yellowish or tan layers of the Kaibab Formation, marine limestone also of Permian age. Above these rocks is the dark red

The oldest rocks in the park—the Permian Cutler Sandstone and Kaibab Limestone—are visible in canyon depths at the Goosenecks of Sulphur Creek. Ray Strauss photo.

At Chimney Rock, unusually rapid erosion creates vertical walls of red Moenkopi Formation mudstone, normally a slope-former. Highest cap is resistant Shinarump Conglomerate. Ray Strauss photo.

Moenkopi Formation, on which the viewpoints are sited.

The Goosenecks are entrenched meanders established when Sulphur Creek flowed with a gentle gradient across loose, poorly consolidated gravel deposited here in Pleistocene time. As the river deepened its canyon through Capitol Reef, it retained its old meanders.

Red sandstone and siltstone blocks lie scattered on the surface here, tilted in every direction as softer mudstone layers beneath are washed or blown away. Because two sets of joints, at right angles to each other, cut through these rocks, many of the blocks are almost rectangular. Some are ornamented with 200-million-year-old ripple marks; some show polygonal patterns of mud cracks, small craterlike raindrop pits, or cube-shaped casts of salt crystals. The lower surfaces of some blocks are marked with unusual angular ridges that may be sand-filled fin marks scraped in soft mud by denizens of Triassic seas.

Grand Wash. In its western part Grand Wash is walled with 250-meter (800-foot) cliffs of Wingate Sandstone. Farther east, where this formation is below the surface, the canyon cuts into Navajo Sandstone. Both formations are patterned with subtly colored mineral matter, dotted with lichens, streaked with desert varnish. Vertical joints effectively maintain the steepness of canyon walls.

At the Grand Wash Narrows the canyon walls, only 6 meters (20 feet) apart at their base, are 150 meters (500 feet) high. The wash goes all the way through Capitol Reef, and makes a nice walk if arrangements can be made for a pickup on the other side.

Hickman Natural Bridge. This natural bridge was formed by gradual deepening of a waterpocket, which eventually joined up with a natural alcove in the cliff below. Weathering and further erosion enlarged the space below the bridge. The bridge is in the Kayenta Formation.

The trail to Hickman Bridge climbs through Wingate Sandstone, then leads across slickrock slopes. Many erosional forms that characterize Capitol Reef can be seen close at hand from this trail: rounded highlands marked with whorls of cross-bedding, barren slickrock slopes merging downward with ever-steepening cliffs, deeply etched overhangs shaped into small natural bridges. Capitol Dome and its neighboring summits rise nearby.

Notom Road. Leaving Utah Highway 24 just east of the national park, this route runs south along the east side of Waterpocket Fold. The road passes pinnacles of Entrada mudstone, then follows valleys eroded in tilted Jurassic and Cretaceous strata, curving in a gentle S parallel to the curve of Waterpocket Fold. To the west, Capitol Reef contains hidden waterpockets and natural bridges. To the east, a stockadelike ridge of Dakota Sandstone crowns colorful Bentonite Hills in the Morrison Formation. Resistant gravel and sand beds in the Morrison Formation contain petrified wood.

The Henry Mountains to the east, though outside the national park, are very much part of the geologic scene on this side of Capitol Reef. They consist of a cluster of laccoliths, igneous intrusions that pushed overlying sedimentary layers up into a dome. Except around the mountain base, the sedimentary rocks that once covered them are now eroded away. The ridge below the Henry Mountains exposes Cretaceous rocks: gray, slope-forming Mancos Shale and above it the Mesaverde Group.

South of Cedar Mesa, the road rises to the Bitter Creek Divide, and south of there it closely parallels the Dakota Sandstone ridge, where fossil oyster shells are abundant. (No collecting within the national park, please.) Navajo Sandstone exposures to the east are marked with caves and alcoves, cross-bedding whorls, and many ravines and gorges. Dark red rock along their base is the Carmel Formation, above the Navajo Sandstone.

The Burr Trail branches from this route north of The Post. It offers first-class views of Waterpocket Fold and the valleys parallel to it. The route also gives access to Muley Twist Canyon and high parts of the fold.

Scenic Drive. Winding through dark red-brown sandstone, siltstone, and mudstone of the Moenkopi Formation, this drive skirts the base of the Wingate cliffs and gives access to Grand Wash and Capitol Gorge. Along the road the Moenkopi Formation is spangled with shiny slabs of selenite, a form of gypsum, broken from veins that lace this rock. Elsewhere, red sandstone and siltstone are decorated with ripple marks and mud cracks.

East of the road, the Chinle Formation appears in purplish gray slopes ridged by thin, hard ledges of conglomerate. As it erodes it undermines the Wingate cliffs. Large slabs fallen from the 300-meter (1000-foot) wall may be marked with broad wind-formed ripple marks. Regulated by joint patterns, the cliff in places is decoratively fluted.

Farther south, the Moenkopi and Chinle Formations are separated by a resistant but discontinuous conglomerate layer, the Shinarump Conglomerate, considered part of the Chinle Formation. Erosion has carved many unusual shapes in this steeply tilted, pebble-filled rock layer.

OTHER READING

Olson, Virgil J., and Muench, D., 1972. *Capitol Reef—the Story behind the Scenery*. KC Publications, Las Vegas, Nevada.

Roylance, Ward J., 1979. *Seeing Capitol Reef National Park, a Guide to the Roads and Trails*. Wasatch Publishers, Salt Lake City.

Chaco Culture National Historical Park

Established: 1907 as a national monument, 1980 as a national historical park
Size: 138 square kilometers (53 square miles)
Elevation: About 1830 meters (6000 feet) at visitor center
Address: Star Route 4, Box 6500, Bloomfield, New Mexico 87413

STAR FEATURES

• The ruins of 13 prehistoric pueblos in or near Chaco Canyon, where geology, archaeology, solar astronomy, and climatology come together to define a widespread and highly developed prehistoric culture.

• Cliffs of Cretaceous rock, well exposed and with weathering characteristics typical of a desert environment.

• A stream valley whose deposits reveal a long history of cyclic erosion and valley filling. Late stages of the erosion cycles had a pronounced impact on local inhabitants.

• Visitor center, museum, introductory movie, evening programs and conducted tours, trails with guide leaflets describing the ruins.

SETTING THE STAGE

In and near flat-floored, cliff-walled Chaco Canyon are sites of numerous pueblos constructed

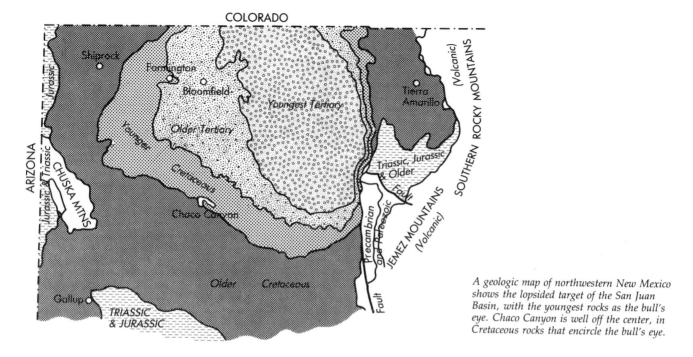

A geologic map of northwestern New Mexico shows the lopsided target of the San Juan Basin, with the youngest rocks as the bull's eye. Chaco Canyon is well off the center, in Cretaceous rocks that encircle the bull's eye.

between 900 and 1100 A.D. by agricultural peoples whom archaeologists call the Anasazi. The canyon is walled with the Cliffhouse Sandstone, part of the Mesaverde Group—a Cretaceous rock deposited 70 to 80 million years ago near the shore of a shallow sea. The rock contains fossil shellfish, coal, shale, clay, and gypsum—all of which, with the sandstone used for building, played parts in the lives of the early inhabitants.

The large pueblos that occupy a 20-mile stretch of Chaco Canyon were apparently the center of a complex Anasazi culture involving at least 59 pueblos, connected by 200 miles of roads. Fine stonework, black-and-white pottery made of local clays, and the use of coal, shale, and gypsum in jewelry and ornaments characterize the pueblo people. Chaco's inhabitants, moreover, established a unique "sun dagger," with a slender sliver of sunshine illuminating spirals carved in a rock cliff on the face of Fajada Butte—an accurate and still functional indicator of solstices and equinoxes.

Though the San Juan Basin, the wide desert in which the ruins occur, is lower than the plateaus to the west, it nevertheless is part of the great stable raft of the Colorado Plateau. The area drains northward, toward the San Juan River, but is given the name "basin" because sedimentary layers sag downward toward its center.

Chaco Canyon is floored with soft gray silt, visible in the walls of the present Chaco Arroyo, and with windblown sand, some of which covered the ruins of the Anasazi villages and sheltered them

from centuries of erosion. Despite this protection, the ruins show the effects of natural processes: gradual disintegration of wooden roof beams and mud mortar used in building, partial destruction by earthquakes and rockfalls, the tumbling of a great block of sandstone near Pueblo Bonito (despite early attempts to stabilize it).

A worrisome aspect of the wide distribution of Chaco ruins, many of which lie outside the national park, is that the San Juan Basin contains one-sixth of the world's known uranium reserves, one-fourth of the near-surface coal deposits of the United States, and plentiful oil and gas. Development of these resources will involve bulldozing of roads, accelerated water use (which lowers the water table and leads to increased erosion), the vibrations and destruction of blasting, and, of course, an influx of people—with concurrent pot hunting and vandalism.

GEOLOGIC HISTORY

Mesozoic Era. Pre-Cretaceous history of this region is probably similar to that of the rest of the Plateau country—repeated incursions of a western sea during the Paleozoic Era, and mostly nonmarine sedimentation in Triassic and Jurassic time. In Cretaceous time the sea once more advanced across the land, coming this time from the east. The advancing and retreating sea deposited a thick shale and sandstone sequence now divided into separate formations, as shown in the accompanying diagram.

The youngest formation of the Mesaverde

Pueblo Bonito was built on the floodplain of Chaco Wash, within easy reach of water for farming. The wash now flows in the deep gully in the middle distance.

Group, the Cliffhouse Sandstone, forms the walls of Chaco Canyon. A near-shore sandstone deposited as the Mesaverde sea retreated eastward, the unit represents the varied environment one would expect along a shore: In addition to an abundance of sand, it contains coal, shale, clay, and gypsum. Above it, toward the center of the San Juan Basin, are formations representing two subsequent advances and retreats of the Cretaceous sea. At the end of Cretaceous time the land

⌐ose once more, and ever since has been above sea level.

Cenozoic Era. Tertiary deposits to the east reflect uplift of the Rocky Mountains at the end of Cretaceous time. Downward warping of the San Juan Basin occurred at about the same time, with regional uplift about 10 million years ago boosting the basin to its present 2000-meter (6000-foot) elevation.

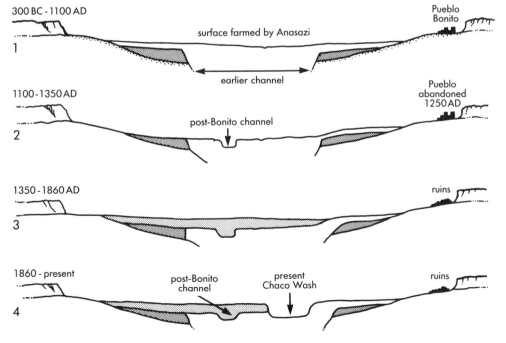

300 BC - 1100 AD

Pueblo Bonito

surface farmed by Anasazi

1

earlier channel

1100 - 1350 AD

Pueblo abandoned 1250 AD

post-Bonito channel

2

1350 - 1860 AD

ruins

3

1860 - present

post-Bonito channel

present Chaco Wash

ruins

4

Chaco Canyon has seen several cycles of filling and trenching. Trenching of Chaco Wash may have caused abandonment of Pueblo Bonito and other Anasazi villages in 1250 A.D. Sections are not to scale.

Chaco Canyon was probably carved in Pleistocene time, when glaciation to the north, increased rainfall in this area, and downcutting by the San Juan River sped up erosion. Within the canyon the oldest deposits are late Pleistocene gravel. Studies of silty stream deposits above the gravel, and of fossil pollen that they contain, show that there have been several episodes of valley filling, separated by episodes of erosion, since Pleistocene time.

Pollen studies show an arid cycle from 5600 to 2400 B.C., when pine and pinyon forests that had earlier covered this area were greatly diminished. Pinyon woodlands began to re-establish themselves—reflecting increased rainfall—around 850 years ago.

When the great pueblos were occupied, between 900 and 1250 A.D., the climate was even drier and warmer than it is today. The Chaco Arroyo was probably a shallow streambed bordered by willows, sedges, and cottonwood trees. The pueblo inhabitants developed an elaborate water-control system that involved dams, canals, and reservoirs, with stone-lined irrigation ditches bringing water to farmed terraces. The system used both stream water from Chaco Creek and runoff from smaller creeks draining surrounding cliffs.

Why did the Anasazi, an obviously thriving people, abandon their pueblos? Poor land use may have played a part in the tragedy. And around 1200 A.D., subtle climate changes brought an increase in erosion. The shallow streambed deepened by as much as 3 meters (10 feet), putting its waters out of reach of the irrigation system. Within 50 years the pueblos were abandoned. Within another 50 years the channel began to refill with sand, the pinyon woodlands again expanded, and the climate became similar to that of today—probably more livable than it was at the height of Anasazi occupancy.

Channeling of the present Chaco Arroyo dates from about 1860, and is part of a cycle of gullying recognized throughout the Southwest. Here at Chaco, erosion control initiated in 1935 to protect the ruins is responsible for accumulation of as much as 2 meters (6 feet) of new sediment in Chaco Arroyo.

OTHER READING

Armstrong, Ruth W., 1980. "The Mystery of Chaco Canyon." *National Parks and Conservation Magazine*, vol. 54, no. 8, pp. 20-24.

Colorado National Monument

Established: 1911
Size: 83 square kilometers (32 square miles)
Elevation: 1408 to 2166 meters (4620 to 7107 feet)
Address: Fruita, Colorado 81521

STAR FEATURES

• Soaring cliffs and imprisoned canyons carved in rock layers of the Uncompahgre Plateau, a fault block lifted 2000 meters (6700 feet) or more above its former position.

• Precambrian metamorphic rock 1 to 2 billion years old, part of the faulted core of the uplift, and overlying Mesozoic strata that record a succession of seas, swamps, desert dunes, lush floodplains, and long periods of erosion.

• Nearby evidence of the strange drainage history of the Colorado River.

• Nature walks, self-guiding trails, and evening programs, as well as roadside exhibits along scenic Rim Rock Drive.

SETTING THE STAGE

This national monument is at the northeast edge of the Uncompahgre Plateau, a fault block 200 kilometers long and 50 kilometers wide (125 by 30 miles) whose core is made of the same hard, dark Precambrian gneiss that forms the forbidding cliffs of Black Canyon of the Gunnison and the Inner Gorge of Grand Canyon. Above this ancient rock lie horizontally layered sedimentary rocks typical of the Colorado Plateau. Formations exposed here extend into adjacent states and appear in other national parks and monuments of the Colorado Plateau.

Landforms here are characteristic of regions where sedimentary strata are thick and horizontal and rainfall is relatively scarce. Because vegetation is sparse, erosion, largely by storm runoff and wind, has a free hand. And because resistant sandstone layers alternate with less resistant mudstone and siltstone, dizzying cliffs and sharply defined ledges alternate with fluted slopes and benches. Cliff height is dictated by the thickness and massiveness of sandstone layers, cliff verticality by their tendency to break away along vertical fractures.

With every storm, cliffs are further undermined and canyons bite deeper into the plateau. Through the centuries, parts of the plateau are little by little cut off, to become freestanding mesas. As erosion continues, mesas are narrowed into buttes and buttes into pinnacles, which eventually erode away completely. Streams, dry most of the year, show branching patterns that are governed to some extent by joints in the cliff-forming sandstone. Erosion along joints defines many features of the park: the Coke Ovens, Sentinel Spire, Balanced Rock, and others.

Many rock surfaces wear a thin black or dark brown coat of desert varnish. Others are marked with lichens.

Cross-bedded Triassic sandstone of the Wingate Formation forms the cliffs of Colorado National Monument and such monoliths as Independence Monument. Ray Strauss photo.

The large fault along the northeast side of the Uncompahgre Plateau is typical of many Plateau country faults. While age-hardened Precambrian rocks that underlie the plateau were broken by the fault, sedimentary rocks above them stretched and draped over the edge of the fault like a giant bedspread, forming a prominent monocline. This downward bend of sedimentary strata along the flank of the plateau can be seen from many places along Rim Rock Drive.

GEOLOGIC HISTORY

Precambrian Era. Dark reddish Precambrian gneiss and schist floor deep canyons of the monument and form the lowest scrub-covered slopes. They are representative of the highly distorted rock that underlies most of the continent. Once sedimentary or volcanic rock, they are now altered beyond recognition. The only datable events in their history are their last recrystallization about 2.3 billion years ago, and their intrusion by granite and pegmatite veins about 1.3 billion years ago.

The upper surface of the Precambrian rock appears as a straight horizontal line below the lowest red sediments. This surface represents an immensely long period of erosion, possibly as long as a billion years, during which whole mountain ranges were slowly stripped away. Final smoothing may have been accomplished by the sea, with the level surface forming just at wave base, where turbulence and wave action reduce rocks to a remarkably even plain.

Paleozoic Era. As this ancient surface sank far-

ther beneath the sea, layers of marine sedimentary rocks were deposited. Between 570 million and 300 million years ago, nearly 700 meters (2000 feet) of marine sandstone, shale, and limestone accumulated, much of it richly endowed with shells of marine organisms. We know the history of these rocks—the advances and retreats of the sea, the rise and fall of nearby land sources—only from adjacent areas, for no Paleozoic strata remain in the national monument.

In the middle of Pennsylvanian time, several large islands rose in roughly the position of the present Rockies. The westernmost of these islands geologists call Uncompahgria because it occupied about the same position as the present Uncompahgre Plateau. Exposed to the elements through the remainder of the Paleozoic Era and part of the Mesozoic Era, the marine sedimentary layers of Uncompahgria, accumulation of 270 million years, were cleaned off, right down to the old Precambrian surface.

Mesozoic Era. In the Uncompahgre region, Mesozoic deposition began soon after Paleozoic strata were eroded away, so the Precambrian surface as we see it today probably corresponds fairly closely to the eroded surface formed at the end of Precambrian time. As the Mesozoic Era began, this surface was barely above sea level. Red sand and mud of a delta or floodplain spread across it. These sediments, now the Chinle Formation, appear near both monument entrance roads—the red rocks just above the Precambrian gneiss.

Later, a creeping sea of windblown sand gradually covered the area. The climate was arid, and the scene definitely Saharan. The dune sands are

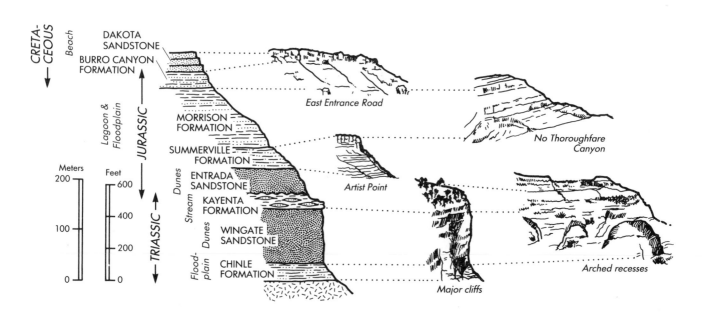

CRETACEOUS

Beach

DAKOTA SANDSTONE

BURRO CANYON FORMATION

JURASSIC

Lagoon & Floodplain

MORRISON FORMATION

SUMMERVILLE FORMATION

Meters 200

Feet 600

ENTRADA SANDSTONE

Dunes

400

KAYENTA FORMATION

100

Stream

TRIASSIC

200

Dunes

WINGATE SANDSTONE

Flood-plain

CHINLE FORMATION

0

0

East Entrance Road

No Thoroughfare Canyon

Artist Point

Major cliffs

Arched recesses

Jurassic and Triassic strata bend in a monocline across the fault that edges the Uncompahgre Plateau's east side.

now the Wingate Sandstone, over 100 meters (330 feet) thick and the major cliff-former of the national monument. Visible on exposed surfaces are the long, sloping laminae of the former dunes. Some surfaces are marked with ripple marks, testifying to winds that swirled and eddied among the dunes 200 million years ago.

The Wingate cliffs are capped by a relatively thin layer of light-colored Jurassic sandstone and conglomerate, alternating with red and purple mudstone: the Kayenta Formation. These rocks were deposited in an increasingly moist climate as streams began to weave among and across the desert sands. Well cemented with silica, the lowest sandstone of this unit acts as a hard caprock protecting the high Wingate Sandstone cliffs. Where the caprock has been stripped away, unprotected Wingate Sandstone weathers into rounded domes, as at the Coke Ovens. The Ka-

yenta Formation also forms the bench followed by parts of Rim Rock Drive.

Above the Kayenta Formation there is another gap in the record, expressive of continued sporadic uplift of Uncompahgria. Two Jurassic rock units known in adjacent regions, the Navajo and Carmel Sandstones, are missing here. Instead, clearly visible beside the Rim Rock Drive north and south of the visitor center, is the record of another invasion of dune sand: thick, broadly cross-bedded, salmon-colored sandstone of the Entrada Formation, which forms a smooth-surfaced upper cliff. Horizontal layers within this rock are interdune deposits, silt and clay that accumulated in open spaces between individual dunes. Some horizontal beds represent short-lived interdune ponds.

At the time the Entrada dunes marched across the land, a shallow sea lay not far to the west, so

The Wingate Sandstone breaks away along vertical joints, as at the right, or, where joints are fewer, in soaring arches like that at the center. Foreground slopes and ledges are in dark "redbeds" of the Chinle Formation. Ray Strauss photo.

these are, in reality, coastal dunes. An arm of the sea, isolated from the main body of water by sandbars, gradually reached into the monument area, and in that shallow bay sand and silt and mud accumulated. The lagoon sediments have become the sandstone, siltstone, and mudstone of the Summerville Formation, seen in the slope just above the Entrada cliff. Ripple marks on the flat flagstone layers suggest wave or current action. Thinner, smoother beds suggest deposition in quiet water.

Eventually, a new sequence of floodplain deposits accumulated, colorful gray and green and red siltstones now known as the Morrison Formation. This unit points to a moister climate, development of shallow lakes and ponds, a rich supply of vegetation, and a correspondingly rich fauna. Many dinosaur skeletons have been found in the formation, including a monster that stood 20 meters (60 feet) high, discovered on the Uncompahgre Plateau south of the national monument. In addition to the skeletons, the Morrison Formation contains highly polished gastroliths, gizzard stones essential to dinosaur digestion. At the other end of the size scale, shells of freshwater snails and clams, and strange little water plants called charophytes, occur in the Morrison Formation, too.

Colorful siltstones of the Morrison Formation cover much of the plateau surface, and appear as well along the entrance roads and at the base of the steep eastern scarp, where the upturned strata are carved into barren, bright-hued badlands.

Late in Jurassic time and early in Cretaceous time, rivers and streams deposited more sand and silt here, mixed with conglomerate derived from rising mountains to the west. These sediments, now the Burro Canyon Formation and the Dakota Sandstone, cap some of the higher parts of the national monument. They, too, are visible along entrance roads near the monument boundary. The Dakota Sandstone, which forms a prominent hogback there, collected along the shores of a Cretaceous sea that entered the region from the east. It is a thin but widespread layer of light tan sandstone that in places contains pebble zones and layers of black, coaly shale.

Cenozoic Era. Rocks younger than the Dakota Sandstone, seen in gray slopes and palisades across Grand Valley, were eroded off the Uncompahgre Plateau as it rose to its present elevation. They show that during late Cretaceous and early Tertiary time, sedimentation rates vastly increased as the Rocky Mountains rose. About 2000 meters (6500 feet) of Cretaceous marine shale and sandstone underlie Grand Valley and form the gray and yellow slopes that culminate in Book Cliffs, the northern and northeastern boundary of Grand Valley. Above Book Cliffs are 2700 meters (8800 feet) of Tertiary deposits, products of floodplain, delta, and lake environments, that culminate in Roan Cliffs. Both Cretaceous and Tertiary sediments once extended across the region that is now the Uncompahgre Plateau.

It is hard to say exactly when the present Uncompahgre Plateau came into existence. As the Rockies began to rise, it certainly was not here at all. South of Colorado National Monument a large, straight canyon cuts through the plateau, deep into its hard metamorphic core, suggesting that a powerful river, probably the ancestor of the present Colorado, flowed across this area as if no plateau were present. Then, imprisoned in its own valley and unable to change its course, it was forced to keep channeling through the rising plateau in Miocene and Pliocene time, eroding through the sedimentary layers and deep into hard Precambrian rocks at the core of the plateau. The river finally did abandon its course, but only when its flow was pirated by a smaller stream that had channeled very rapidly through the soft Cretaceous and Tertiary siltstone in Grand Valley. Today the Colorado follows the course of that smaller stream northwest around the north end of the Uncompahgre Plateau, leaving what may have been its former channel as Unaweep Canyon.

OTHER READING

Lohman, S. W., 1965. *The Geologic Story of Colorado National Monument*. Colorado and Black Canyon Natural History Association, Grand Junction, Colorado.

U.S. Geological Survey, 1963. *Geological Map on Grand Junction Area, Colorado*. Miscellaneous Investigations Map I-404.

El Morro National Monument

Established: 1906
Size: 5 square kilometers (2 square miles)
Elevation: 2380 meters (7218 feet) at visitor center
Address: Ramah, New Mexico 87321

STAR FEATURES

• Petroglyphs and inscription-carved cliffs of Jurassic sandstone formed from ancient dunes.

• An unusual but dependable waterhole that for centuries has furnished plentiful water in an otherwise arid land.

• Erosional landforms that include Inscription Rock and nearby cliffs, with clefts and alcoves shaped by water, wind, frost, and gravity.

• Visitor center displays, evening programs, and a descriptive trail leaflet for walks around and over Inscription Rock.

SETTING THE STAGE

The Zuni Mountains of western New Mexico, an oval, dome-shaped uplift, have a central core of Precambrian granite, gneiss, and schist. Tilted layers of sedimentary rock that once blanketed the whole area now appear in concentric bands around the denuded core, the resistant ones shaping hogbacks and cuestas whose cliffs stand like rings of bleachers, the weak ones eroded into

racetrack valleys between. This pattern, though easy to see from the air or on a geologic map, is obscured by its very size when seen from the ground, especially since parts of the sedimentary bands are covered with lava flows. Prominent among the outermost rings is the Zuni Sandstone, curving around the southeastern end of the uplift and forming the cliffs of El Morro National Monument.

This sandstone, fairly soft and easily carved, capped by protective ledges of harder Dakota Sandstone, rises above a valley of weak but colorful shales of the Chinle Formation. Its long, sweeping cross-bedding and fine, evenly rounded, frosted quartz grains show that the Zuni Sandstone is a dune deposit. The darker, browner Dakota Sandstone above it is a mixture of pebbles and sand probably deposited by streams flowing across a coastal plain near the sea.

Inscription Rock and the bluffs near it (*el morro* means a headland or bluff) reveal many features of both of these sandstones, and show weathering characteristics typical of semiarid lands. Along the trail at the base of the cliff, and on the trail over the top of the mesa, you will see examples of these processes. Watch for:

• Desert varnish, formed as water seeps from the rock face and evaporates, leaving on the surface tiny amounts of manganese and iron minerals supplied by desert dust or dissolved and carried from the interior of the rock.

• Arched alcoves that develop as blocks of unsupported rock break away and tumble from the cliff face.

• Deep horizontal clefts caused as descending groundwater flows horizontally along relatively impermeable layers of fine clay that mark the position of old interdune deposits.

• Vertical faces that result from weakening, breakage, and erosion along prominent vertical joints. In places the broken faces show arches of curving conchoidal (shell-like) fractures. Elsewhere they display exfoliation, the peeling off of thin scabs of rock in a continuous weathering process.

• Details of the uneven contact between the Zuni Sandstone and overlying Dakota Sandstone, a contact that represents a 30-million-year break in sedimentation. Channels along the contact are commonly filled with wave-washed blocks of white Zuni Sandstone imbedded in the browner shoreline sands of the Dakota Formation.

• On the trail over the mesa, views into Box Canyon—cliff-walled and remarkably deep for a canyon with such a small drainage area. Joints in the Zuni Sandstone, as well as the varying hard-

Names and dates are carved in soft Zuni Sandstone, formed in Jurassic dunes.

The plunge pool at the base of El Morro's cliffs collects runoff from the mesa surface. Streaks of black lichens and desert varnish mark seepage channels in the rocks above.

ness of the sandstone, seem to have provided an avenue for unusually rapid erosion.

* * * * *

The waterhole at the base of Inscription Rock was no doubt the main attraction here in prehistoric days as well as in the days of Spanish and Anglo-American exploration and settlement. It is not a spring, but a deep pool filled by rain and snowmelt draining from the mesa top down deep joint clefts in the sandstone. The names on the rock above the pool were carved at a time when a sandy beach extended along one side. After a rockfall filled the waterhole in 1942 (leaving a light-colored, relatively unweathered rock face to show where it broke away) both rock and sandbar were removed. The small dam is a modern addition, slightly increasing the size of the pool.

GEOLOGIC HISTORY

The geologic story here is no less interesting than the inscriptions on Inscription Rock, but very little of it can be deduced from El Morro alone. We must look to the rest of the Zuni Mountains, and indeed to the entire Plateau region, for the rest.

Precambrian and Paleozoic Eras. The granite, gneiss, and schist exposed in the Zuni Mountains tell of ancient mountain building, and surrounding rings of sedimentary rock tell us that during the 330 million years of the Paleozoic Era the area was covered by a succession of seas coming in from the west. Flat-lying marine limestone, sandstone, and shale layers deposited in these seas are similar to those at Grand Canyon National Park, but here, periodic uplift caused many layers to be stripped away almost as soon as they were deposited. Only Pennsylvanian and Permian strata were preserved, lying directly on Precambrian rocks. These strata correlate with the uppermost layers at Grand Canyon and have counterparts in the cavernous limestone of Carlsbad Caverns, equally far away in the other direction.

Mesozoic Era. As the Mesozoic Era opened, the area lifted and the sea retreated once again, though the land remained low-lying like the coastal plain of the Gulf Coast today. In lagoons, marshes, and shallow rivers, and on gently shelving shores of Triassic time, thick layers of silt and mud were deposited. In Jurassic time these strata were covered over with dunes of windblown sand that eventually became the Zuni Sandstone. In Cretaceous time they were covered again, by the brownish sand and mud of the Dakota Formation.

Thin but extremely widespread, the Dakota appears in many western states. It signals a new advance of the sea, this time from the east and southeast. As this sea spread westward, layers of dark gray marine shale were deposited in it throughout the Colorado Plateau and Rocky Mountain regions. Later, as the sea retreated, shore sands appeared again, thicker this time and with interlayered beds of plant material that later became coal. Then, near the end of Cretaceous time, about 70 million years ago, the land lifted and the sea drew back "for good." (Geologists, with their long-term view, know it will someday come again.)

Cenozoic Era. In New Mexico the early part of the Cenozoic Era was a time of uplift, volcanism, erosion, and more volcanism. As the Zuni Mountains pushed up, erosion carved deeply into the sedimentary rocks that covered them. Volcanoes showered ash upon the land, and molten rivers of lava crept down stream valleys and across the eroded surface. Eventually the entire region, with all its valleys and mountains, was lifted to its present elevation. Uplift aided and abetted erosive forces and brought about renewal of volcanism. Rain and wind, helped by frost and melting snow of the Pleistocene Epoch's cooler, wetter climate, deepened canyons and valleys, sharpened cliffs and ridges. Desert winds and sporadic storm-fed torrents continue these processes today.

Within the last few seconds of geologic time, prehistoric peoples, Spanish conquistadores, and Anglo-American explorers, soldiers, and settlers carved their names and the dates of their passing on the tablet of cliffs around the waterhole.

OTHER READING

Anonymous. *El Morro Trails*. Southwest Parks and Monuments Association, Globe, Arizona.

Foster, Roy W., 1971. *Southern Zuni Mountains: Zuni-Cibola Trail*. Scenic Trips to the Geologic Past, no. 4. New Mexico Bureau of Mines and Mineral Resources, Socorro, New Mexico.

Grand Canyon National Park

Established: 1908 as a national monument, 1919 as a national park
Size: 4929 square kilometers (1904 square miles)
Elevation: 613 to 2792 meters (1800 to 9161 feet)
Address: P.O. Box 129, Grand Canyon, Arizona 86023

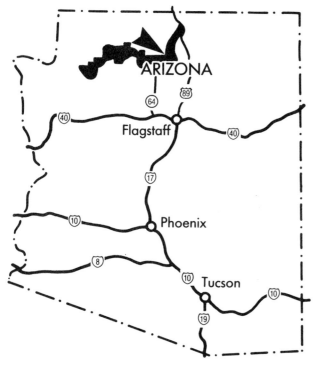

STAR FEATURES

• Geology's masterpiece, an awesome abyss unsurpassed for scenic grandeur. Grand Canyon's walls, with their orderly array of sedimentary layers lying horizontally above a beveled surface of much older rocks, unfold the pages of 2 billion years of Earth history.

• Erosional landforms geologically young and typical of regions of horizontal strata and desert climate.

• Four hundred fifty kilometers (280 miles) of the Colorado River, the master sculptor, itself ruled by uplift in distant plateaus and mountains.

• Volcanic features that add dark overtones in western Grand Canyon.

• Many trails, some with trailside signs interpreting the geology. Rim routes are pleasant and easy; trails into the canyon are strenuous. A trip to the river on foot or by mule is a trip through time, far into the distant past.

• Roadside displays, nature walks, interpretive talks, shuttle-bus tours (summer only) to viewpoints along the rim, and a visitor center and museum touching heavily on the history of geologic exploration of the canyon.

• At Tusayan, just outside the park, an IMAX film portraying human history and exploration of Grand Canyon, with an armchair view of the exciting world of river running.

See color pages for additional photographs.

From the dark Precambrian gneiss of the Inner Gorge (lower left) to the white Permian limestone of the canyon rim, Grand Canyon offers an unparalleled view of much of the Earth's history.

SETTING THE STAGE

One of the marvels of Grand Canyon is that it is here at all. Why should a river develop a course directly across a 3000-meter (9000-foot) uplift? For the Colorado River bisects the very heart of the Kaibab Uplift, and bisects it so thoroughly that the river flows in desert summer while pine and aspen clothe the cool plateaus above, so thoroughly that animals south of the canyon have evolved differently than those north of the canyon, so thoroughly that it is an effective barrier even to man, who must drive 380 kilometers (235 miles) to see the canyon from the other side.

The great chasm trends roughly east-west. Because the surface of the uplift and the rock layers within it slope gently southward, regional drainage on the North Rim flows *toward* the canyon, while that on the South Rim flows *away* from the canyon. The largest tributaries therefore have established themselves north of the river—Bright Angel, Shinumo, Kanab, and other creeks—and the Colorado River is much closer to the South Rim than to the North Rim. The river is 1360 meters (4460 feet) below Grand Canyon Village on the South Rim, and 1810 meters (5940 feet) below Grand Canyon Lodge on the North Rim.

Dark, resistant metamorphic rocks—the Vishnu Schist—wall Grand Canyon's somber Inner Gorge. Ray Strauss photo.

Through the ages the Colorado River has cut downward through one rock layer after another, in a peculiar history that involves drainage patterns in Arizona, Utah, and Colorado. But its once turbulent waters are tamed now by dams that retain floods and withhold the sand, gravel, and rock that used to come from far upstream. With this change, deepening of the canyon has now ended, though widening by tributary streams and rockslides continues.

Through much of the canyon's length the river is imprisoned by resistant rock of the narrow Inner Gorge. It cannot swing widely to open up a broad valley, as do most major rivers. Canyon widening we must credit to other agents: storm-fed tributary streams, temperature changes, wind, and gravity. Summer storms, severe but brief, bring every ravine, every gulch to life, and churning rivulets plunge from cliff to cliff until the canyon is a world of waterfalls, many of them red

with silt and mud from redrock slopes. The short-lived torrents carry silt, sand, and pebbles into tributary canyons. Reaching at last the lowest cliffs, they cascade into the Colorado and contribute their colorful loads to the river. The largest storms fill main tributary canyons with tumbling torrents that tear loose and carry rock fragments of all sizes downstream toward the river.

Daily temperature changes in the canyon are sufficient to crack the rocks of its walls. On frosty nights, frequent enough along canyon rims but less common in its depths, frost forming in cracks and crevices wedges rocks apart. Angular slabs, some tiny, some tremendous, are by degrees pried away, and finally, perhaps only after many centuries of such frost action, they tumble and slide to the slopes below. Rockfalls and landslides widen the canyon under the influence of gravity. Those who have lived on the rim have heard their thunder; piles of angular boulders or patches of

Grand Canyon's walls reveal an orderly sequence of Paleozoic sedimentary rocks, from the Cambrian Tapeats Sandstone to the Permian Kaibab Formation.

newly exposed, unweathered rock tell their story. Many rockfalls are caused by oversteepening of slopes and undermining of cliffs by wind and rain. Large backward-tilted blocks, common features in parts of the canyon, slid downward along curving fault planes. Some individual buttes in Grand Canyon now stand out from the rim because of fault motion down and away from the rim.

The colorful layers of sedimentary rock show different degrees of resistance to wearing down and widening processes, making of the canyon walls magnificent illustrations of differential erosion. Well-cemented sandstone and limestone tend to break away into vertical cliffs and ledges, their heights corresponding to the thickness of the rock layers. Siltstone and mudstone erode into gentle slopes and benches. In eastern Grand Can-

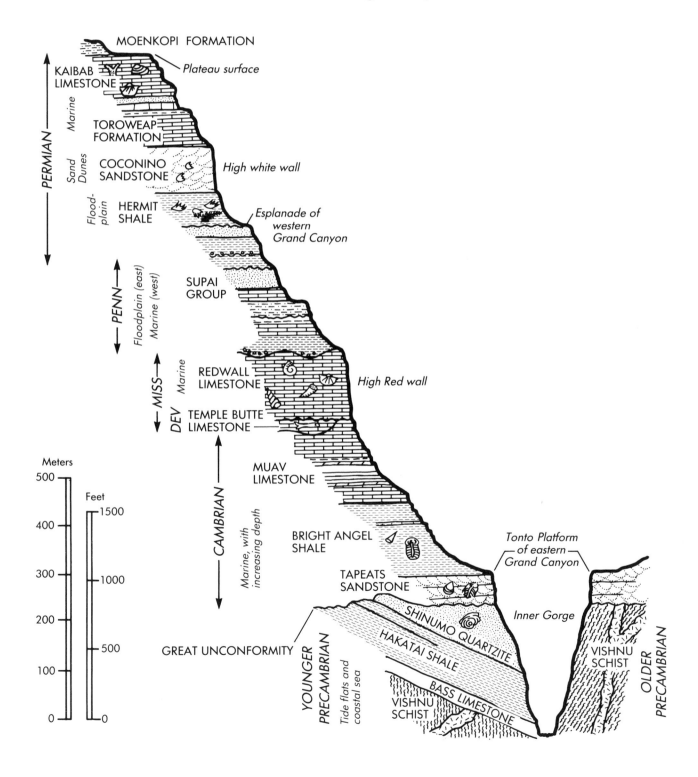

Cliffs of Redwall Limestone and Coconino Sandstone are offset nearly 60 meters (200 feet) by the Bright Angel Fault, which runs down the center of the photograph. The fault forms a route by which the Bright Angel Trail can surmount these cliffs. Ray Strauss photo.

Garden Creek in the foreground and Bright Angel Creek in the distance mark the straight line of the Bright Angel Fault. Indian Gardens, in a clump of cottonwood trees (at arrow), is located where springs rise along the fault. Ray Strauss photo.

yon one bench in particular stands out: the wide Tonto Platform that spreads a gray-green apron above the Inner Gorge. (Tiny trails that cross the Tonto Platform provide a measure of its breadth: They are six feet wide or more.)

Since the sedimentary strata lie in horizontal layers, ledges, cliffs, and benches repeat themselves below every point and across every ravine and canyon, reaching symmetrically toward the dark, hard cliffs of unstratified igneous and metamorphic rocks of the Inner Gorge. Only in western Grand Canyon is there a major change of pattern. There the Tonto Platform disappears, and soft red siltstone and mudstone layers above the Red Wall thicken and erode back into a wide shelf known as the Esplanade.

Despite the symmetry, there are many minor changes from place to place in different parts of the canyon walls, as there are in rocks everywhere. Geologists find these changes particularly exciting here because rock layers can be traced visually for long distances, or followed by walking along well-defined cliff tops or individual ledges. Here, many ideas about geologic processes can be tested. Sandstone interlayered with shale or limestone visibly demonstrates fluctuations in sea level. River-deposited red siltstone giving way westward to shore deposits and eventually to marine limestone demonstrates differences in environment across coastal regions. Ancient channels and even more ancient islands are expressed in surrounding rock layers. Pebble-filled conglomerate on old erosion surfaces indicates the end of erosion and the beginning of deposition. And from bottom to top in the sedimentary layers, fossils reflect the evolution of life from primitive Precambrian algae through shell-bearing denizens of Paleozoic seas, from early fishes to lizardlike reptiles that left their footprints on Permian dunes.

Grand Canyon also furnishes unequalled examples of a primary geologic "law": In undisturbed sedimentary rock layers, the oldest rocks are at the bottom, and those above are sequentially younger and younger. An understanding of this precept adds to our understanding of geologic time and drops each layer of the canyon walls into place: the oldest near the river, the youngest at the rim.

Grand Canyon properly begins where the Little Colorado joins the Colorado, east of Desert View. It ends at Grand Wash Cliffs, the southwest boundary of the Colorado Plateau region. The national park now includes Marble Canyon, upstream from the Little Colorado, and extends west to the stilled waters of Lake Mead. Between the

Little Colorado and the base of Grand Wash Cliffs, plateaus rise and descend stairstep fashion, with faults and monoclinal folds controlling their shapes and height. Grand Canyon shows us that faults and monoclines are closely related: Where Precambrian rocks are broken and displaced, overlying sedimentary rocks bend and drape across the faults. Many of the great faults edging the plateaus originated in Precambrian time and have moved intermittently since then.

GEOLOGIC HISTORY

Precambrian Era. In Grand Canyon, Precambrian rocks fall into three age groups: the highly metamorphosed Vishnu Schist, which had its beginnings more than 2 billion years ago; granite that intruded the schist about 1.75 billion years ago; and a much younger group of sedimentary and volcanic rocks. Studies of the schist show that it was involved in several pulses of mountain building, and altered at depths as great as 20 kilometers (12 miles) below the surface. Intrusion of the granite was accompanied by large-scale faulting, particularly along north-south faults. Mountains formed during metamorphism of the schist and intrusion of the granite were leveled by erosion long before younger sedimentary rocks were deposited.

Precambrian sedimentary and volcanic rocks appear in faulted, tilted wedges in the eastern part of Grand Canyon. Sandstone, shale, limestone, conglomerate, and lava flows are still easy to recognize, even though they were deposited more than a billion years ago. Some are imprinted with ripple marks and cross-bedding; others display relic mudcracks. A few of them contain fossil algae, some of the earliest known traces of life.

After these rocks were deposited they were broken and tilted along earlier faults to form a series of fault-block mountain ranges. These in turn were destroyed by nearly a billion years of erosion—the longest, most widespread period of erosion that the Earth has known. Both older and younger Precambrian rocks were trimmed away to an essentially horizontal surface, a peneplain now expressed by the Great Unconformity between Precambrian and Paleozoic rocks, not only in Grand Canyon but in many other parts of the world.

Paleozoic Era. In Grand Canyon the Great Unconformity is clearly exposed for hundreds of kilometers, and can be studied in greater detail than anywhere else on Earth. Hills and ridges of resistant Precambrian quartzite, islands in the

At the east end of Grand Canyon the Little Colorado River builds a rocky fan. A repeat performance, the delta will be swept away when more water is released through Glen Canyon Dam. Tad Nichols photo.

Cambrian sea, in places protrude upward into Cambrian sedimentary rocks. Imbedded in Cambrian strata are blocks of broken rock torn from the islands by surging waves. Sea cliffs undercut by waves, even the flattened pebbles of shingle beaches, can be found along the Precambrian-Cambrian unconformity.

Cambrian strata here illustrate the detail to which geologic history can be worked out when rock layers are well exposed over long distances. Research shows that the Cambrian sea advanced in five pulses, with predictable changes in envi-

ronment from the shore outward: conglomerate and coarse sandstone near the shore, finer sandstone and mudstone farther out, and finally limestone where the sea was fairly open. Small animals—trilobites, brachiopods, simple cone-shaped gastropods—crawled about on the sea bottom, leaving both trails and shells. Sponges and various forms of seaweed flexed with tidal currents.

Toward the end of the Cambrian Period the sea retreated. There is no record here of Ordovician or Silurian time. Either there were no deposits, or they all eroded away. Shallow channels in the top of the Cambrian layers are filled with thin Devonian limestone—purplish, gnarled, and sugar textured—replaced westward by continuous bands of white and gray dolomite.

After yet another retreat and readvance of the sea, the Redwall Limestone was deposited here in Mississippian time. Thick, strong, and extremely widespread, the limestone contains fossil fish and many small marine shellfish such as brachiopods, corals, mollusks, crinoids (sea lilies), and trilobites. Nodules of chert, a porcelainlike version of quartz, mark horizontal planes in the limestone. The rock is not really red at all, but bluish gray; in Grand Canyon its surface is stained with red iron oxide mud from the colorful shales above.

The top of the Redwall Limestone is marked by erosion similar to that of many limestone areas today, a surface characterized by extremely irregular topography, with many solution caves, sinkholes, and underground channels. This hummocky, irregular karst surface gives us a clue to the climate here in late Mississippian time, for such surfaces are known today in Kentucky, Puerto Rico, Yugoslavia, and other limestone-surfaced regions with warm, humid climates.

During Pennsylvanian and Permian time this region saw more sedimentary deposits. The Supai Group (made up of four formations) and above it the Hermit Shale consist of ledge-forming, cross-bedded sandstone and deep red slope-forming siltstone. Deposited on a floodplain or delta and in an adjoining shallow sea, some of these rocks bear imprints of fernlike leaves; others are marked with reptile tracks. In western Grand Canyon the sandstone beds thicken at the expense of the siltstone; westward-thickening limestone layers with marine fossils show that the sea lay in that direction.

A little later in Permian time, windblown sand swept across the Grand Canyon region, leaving a thick deposit of fine white sandstone marked with long, diagonal cross-bedding—the Coconino Sandstone, the white band on the canyon's upper walls. Tracks and trails of lizardlike animals mark some sloping dune surfaces, but no bones or skeletons of the animals have ever been found.

During the remainder of Permian time, the western sea continued to fluctuate; twice it inundated the Grand Canyon region, depositing the Toroweap Formation and then the Kaibab Formation. These rocks now form stairlike ledges and slopes on the uppermost 150 meters (500 feet) of the canyon walls. Again, Grand Canyon exposures allow detailed studies. The lowest part of each formation shows the sea's advance, with fragments of underlying rock redeposited across an uneven surface. The middle part of each contains limestone with shore-derived sand in eastern Grand Canyon, pure marine limestone in the west. And the upper parts contain layers of reddish silt and beds of gypsum deposited as the sea backed away westward.

Warm and shallow, the Toroweap and Kaibab seas teemed with life, and many fossils can be seen in these rocks: trilobites and brachiopods, snails and clams, cephalopods (shell-bearing cousins of the octopus and squid), feathery bryozoans, corals, and, less commonly, shark teeth.

Mesozoic Era. There are no Mesozoic rocks at Grand Canyon now. However, those in nearby areas show us that not long after the Kaibab sea's retreat, thick layers of sand, mud, and volcanic ash were deposited here. What Grand Canyon has lost can be found in other national parks of the Colorado Plateau: Petrified Forest, Zion, Canyonlands, and Arches. By the end of Mesozoic time, Mesozoic strata covered all the Paleozoic sedimentary rocks.

Cenozoic Era. Late in Mesozoic time and early in the Cenozoic Era, 60 to 50 million years ago, ancient faults were reactivated as this region began to shift and rise. Movement along old north-south faults offset old fault blocks; their blankets of Paleozoic and Mesozoic sedimentary strata stretched and draped across the faults. Many new faults formed, most of them trending northeast on a collision course with the reactivated north-south faults.

Many of the faults show up remarkably well from the air, appearing as steps in the plateau surface or as straight canyons eroded along lines of fault-shattered rock. Such, for instance, is the Bright Angel Fault. Its straight, northeast-trending scar has been widened by Bright Angel Creek in its route from the North Rim to the river, and by Garden Creek from the South Rim to the river. Practically all the major tributaries north of the river follow faults established at this time.

This brings us back to our original question: Why is Grand Canyon here, cutting so boldly across the Kaibab Uplift? The answer is found not in the canyon depths but in the shape of the old uplift, in lava flows, in gravel scattered on plateau surfaces, and in our knowledge of cliff retreat and the tendency of modern streams to flow along the bases of retreating cliffs. Though many questions have not been answered, a sequence of events can

At Toroweap Point in western Grand Canyon, lava flows repeatedly cascaded into Grand Canyon, damming the river. In this area, erosion of Supai Group redbeds forms the wide bench of the Esplanade. Tad Nichols photo.

be reconstructed from these clues:

• Fault movements early in Cenozoic time re-established many Precambrian faults in the Grand Canyon area, divided the various plateaus of this region, and domed up the Kaibab Arch, the broad anticline of the Kaibab and Coconino Plateaus. For some time, streams from this area drained northward down the gentle dip of the plateau strata.

• By Miocene time, as Mesozoic rocks were stripped away, several large rivers draining the Wyoming and Colorado Rockies had joined to

form the Ancestral Colorado River. Flowing southwestward, this river eventually met up with the Kaibab Uplift, at that time still partly covered with Mesozoic rocks.

• The Ancestral Colorado curved southward around the highest part of the uplift, in a broad, irregular curve that followed the base of northward-retreating cliffs of Mesozoic rock.

• Until about 6 million years ago the river may have curved northward or northwestward somewhere near the west end of the present Grand Canyon, to flow into lakes in Utah and Nevada.

You'd think a river as large as the Ancestral Colorado would leave more traces—wide valleys, deep canyons, identifiable river deposits. But faulting, volcanism, and subsidence of intermountain basins west of the Colorado Plateau have so far hidden the ancestral river's exact destination.

• Meantime, around 5.5 to 5 million years ago, well to the south, the Gulf of California opened up—a new rift valley between two branches of the San Andreas Fault. With this new sea-level outlet, streams in that area were strengthened. Some of them combined into a larger, stronger river that established its course along the present Arizona-California boundary, the route of the present lower Colorado River.

• Fortified by additional tributaries, this stream eroded headward, cutting into the southwest edge of the Colorado Plateau—the Grand Wash Cliffs. Breaking through the cliffs that separated it from the Ancestral Colorado River, it "captured" that river and deflected its flow southward to the Gulf of California.

•. Because the gradient of this course was steeper than the old route into Utah or Nevada, the combined rivers—which we can now call the Colorado River—eroded downward rapidly. In less than 5 million years, Grand Canyon as we know it today was excavated along the line established by the Ancestral Colorado as it curved southward around Mesozoic rocks on the highest part of the Kaibab Arch. Farther upstream the Colorado also deepened its canyon, there carving into Mesozoic rock layers to create the canyons of eastern Utah and western Colorado. Tributaries kept pace, chiseling canyons of their own. At the same time, Mesozoic rocks continued to be stripped from southern parts of the Plateau.

• In western Grand Canyon, about 1.2 million years ago, volcanoes erupted, sending lava flows cascading into Grand Canyon. At that time the canyon was only 15 meters (50 feet) less deep than it is now, telling us that most of the carving of Grand Canyon took place before 1.2 million years ago. Erosion has apparently slowed down in the last million years, partly because of those lava flow dams, partly because the river had to cut through hard metamorphic and igneous rocks of the Inner Gorge.

BEHIND THE SCENES

Most of the following areas are in eastern Grand Canyon, the part of the national park that sees the most visitors. The remote Toroweap region of western Grand Canyon, though off the beaten track, is included because of its volcanic features and views of the Esplanade.

Bright Angel Point (North Rim). This narrow point, below Grand Canyon Lodge, offers an excellent view down Bright Angel Canyon—deep and straight because it follows the straight line of the Bright Angel Fault.

Partway along this canyon a spot of bright orange marks the Hakatai Shale, in the younger Precambrian sedimentary sequence. With other Precambrian sedimentary rocks, it forms a wedge beveled at the top by the Great Unconformity.

The Tapeats Sandstone and younger Paleozoic layers lie above the unconformity. From bottom to top they are the Bright Angel Shale, Muav Limestone, Redwall Limestone, Supai Group, Hermit Shale, Coconino Sandstone, Toroweap Formation, and, finally, the Kaibab Formation at the canyon rim. All these rocks can be identified by referring to the stratigraphic diagram.

Bright Angel Trail (South Rim). Zigzagging steeply, the Bright Angel Trail keeps close to the line of the Bright Angel Fault (notice the difference between the height of the rim on either side of the fault), where offset rock layers furnish a handy break in the Coconino Sandstone and Redwall Limestone ramparts. The trail follows the fault almost all the way to the Colorado River, dropping through the entire layer-cake sequence of sedimentary rocks. Trailside signs name the formations. Look in the Kaibab and Toroweap Formations for fossils (no collecting!), in the Coconino Sandstone for dune-style cross-bedding and animal tracks, and in the Hermit and Supai redbeds for ripple marks, leaf impressions, and pebbly layers that establish periods of erosion. The top of the Redwall Limestone shows evidence of solution caverns and rough topography developed before higher layers were deposited.

Not far below the Redwall cliff the trail comes to Indian Gardens, where springs issue from Cambrian rocks to nourish a grove of cottonwood trees. Cambrian Bright Angel Shale borders the trail across the Tonto Platform. At the base of this shale the Tapeats Sandstone forms the ledgy edge of the Inner Gorge. Below the Tapeats is the Great Unconformity, and below that, on the right, the Vishnu Schist. To the left, light-colored Precambrian granite replaces the Vishnu Schist.

Across the river, west of Bright Angel Creek, Precambrian sedimentary rocks lie almost horizontally on top of the Vishnu Schist. They escaped erosion at the end of Precambrian time because they had dropped down between two faults. Rocks exposed along this part of the Inner Gorge are quite severely faulted, with many small faults cutting across the Bright Angel Fault. Many of the rock layers are sharply bent by drag along the faults.

Cape Royal (North Rim). Cape Royal looks out over a tumbled and jumbled sea of rocks that doesn't follow the Grand Canyon pattern of flat rock layers lying neatly one on another. There are three reasons for this: Most of the rocks between Cape Royal and the river are Precambrian sedimentary rocks and lava flows, in wedges tilted in Precambrian time; the Paleozoic rocks themselves curve over the East Kaibab Monocline, the edge of the Kaibab Plateau; many faults complicate the geologic picture here. Some cut and displace only Precambrian rocks; others displace both Precambrian and Paleozoic rocks.

The most easily recognized Precambrian units are bright orange Hakatai Shale and dark Cardenas Lavas, evidence of Precambrian volcanism. Tan surfaces near the river are the Dox Sandstone. By examining the scene carefully (with binoculars if you have them) you can identify these tilted, beveled Precambrian units and the Paleozoic strata lying horizontally across the bevel.

To the southeast, on the other side of the canyon, a red-sloped butte known as Cedar Mountain rises above the surface of the Marble Platform. It is a remnant of Mesozoic sedimentary rocks that once blanketed this region. Farther north, the Little Colorado River, with an impressive gorge of its own, enters the main Colorado River.

Colorado River. Before the Glen Canyon Dam was built, the Colorado in flood carried downstream an average half-million tons of sand, silt, and gravel *per day*, and in addition rolled and bounced many large cobbles and boulders along its bed. The muffled roar of rock hitting rock could often be heard from the rim. Descending 2200 feet in 277 miles, the river at flood stage cleared from its channel much of the bouldery rubble brought in by tributary streams.

Now when such coarse debris is brought to the main channel, the Colorado foams with rapids but can't clear away as much of the rock debris. Without its former "spring cleaning," bouldery deltas build outward from the mouths of large side canyons, constricting the Colorado's channel and enlarging rapids there.

Along the river, rocks are polished and fluted from long abrasion by mud, sand, and gravel. But here, too, we see a slowdown in erosion as the Glen Canyon Dam holds back the river's natural

tools. Without boulders for pounding and gravel for scouring and polishing, the river has little cutting power.

River runners see many fine examples of the river's past work. Float trips begin at Lees Ferry at the upper end of Marble Canyon, at the top of the Kaibab Limestone, and descend gradually through the same rock sequence that occurs on the walls of Grand Canyon. Geologic guidebooks for the river run are listed in the references.

Desert View (South Rim). From this viewpoint, one of the most fascinating in the park, you can see the river below its junction with the Little Colorado, which comes in from the south through a deep, narrow, sheer-walled chasm. To the east is Cedar Mountain, a remnant of Triassic rocks left on the otherwise clean-swept Kaibab Limestone of the Marble Platform. Farther away, dim with distance, are the Jurassic rocks of Echo Cliffs.

The Kaibab Limestone of the Coconino Plateau bends down sharply along the East Kaibab Monocline—almost at your feet—and then levels off again 1000 meters (3000 feet) below as the surface of the Marble Platform. If you came to Grand Canyon from Cameron, you climbed the East Kaibab Monocline just before entering the park.

In the canyon depths, tilted orange, tan, and gray-green rock layers are Precambrian sedimentary strata that survived the long erosion at the end of Precambrian time. The Great Unconformity cuts across them horizontally, and Paleozoic rocks lie across the truncation.

Looking west down Grand Canyon, notice that the North Rim is higher and farther from the river than the South Rim. Before the canyon was here, the Kaibab and Coconino Plateaus together made up a single broad dome, the Kaibab Arch or Uplift, whose highest point was on the Kaibab Plateau, well north of the present canyon. From that high point, the surface of the plateau (as well as its rock layers) slopes gently southward. Because of this, streams north of the canyon flow toward the river, eroding tributary canyons as they go. By the same token, drainage south of the South Rim flows away from the canyon, and takes no part in the canyon-widening process.

Kaibab Trail (North Rim). This trail descends to the river via Roaring Springs and Bright Angel Canyons, both established along major faults. Paleozoic strata can be identified from the stratigraphic diagram.

Roaring Springs contributes to the water supply for Grand Canyon Village on the South Rim.

Prospect Creek's steep-walled canyon, entering from the south (left), marks the position of the Toroweap Fault, as does the lava cascade at the right. Rocky flood-built fans like that at Prospect Creek constrict the Colorado River, forming its major rapids. Ray Strauss photo.

Water that bursts from these cliffside springs comes from rain and snowmelt that sink into the Kaibab Plateau and flow through a network of underground solution channels in Paleozoic limestones. Reaching impermeable layers of shale, the water moves sideways through gently south-dipping strata until it intercepts the canyon walls.

Because of rapid erosion along Roaring Springs and Bright Angel Faults, there is no broad Tonto Platform here. The trail drops rapidly through all the Paleozoic strata. Below the easily recognized, slabby Tapeats Sandstone it reaches the Great Unconformity, the line of contact with Precambrian rocks. If you place your hand across this contact, your fingers span 600 million years of Earth history!

The uppermost Precambrian rocks are sedimen-

tary and volcanic—the Dox Sandstone, Shinumo Quartzite, bright orange-red Hakatai Shale, and Bass Limestone, which you will encounter in this youngest-to-oldest order. At Ribbon Falls, calcium carbonate-laden water plunges over part of a dark diabase sill, depositing its limy load to form a travertine dome. The sill borders the trail for some distance. Farther downstream the trail enters yet older rock, the Vishnu Schist, recognizable by its dark color, shiny crystals, and generally vertical texture. This is the rock of the Inner Gorge, and Bright Angel Creek flows in its own inner gorge most of the way to the river.

The climate change along the trail is pronounced, from a cool Canadian-style forest on the rim to a Sonoran Desert climate at the river. The average temperature difference between rim and

The view northeastward from Desert View shows east-dipping rocks along the East Kaibab Monocline (left), the wide valley carved in faulted Precambrian sedimentary rocks, and the narrow slot of the Little Colorado River's canyon (upper right). In the distance are the Mesozoic rocks of Echo Cliffs. Tad Nichols photo.

river is about 18 degrees Celsius (35 degrees Fahrenheit), making 1.8 vertical kilometers (1 mile) the equivalent of 2000 horizontal kilometers (1200 miles)!

Kaibab Trail (South Rim). Descending from Yaki Point, this trail switchbacks through the Kaibab and Toroweap Formations and the Coconino Sandstone. Many details of these rocks are apparent along the trail: fossils and hard chert nodules in the Kaibab Limestone; red- and yellow-tinted near-shore sandstone interlayered with limestone in the Toroweap Formation; pronounced cross-bedding and wind-ripple marks in the Coconino Sandstone. Notice the fine, even-grained sand weathered from the Coconino Sandstone—typical dune sand.

Below the cliff the trail crosses redbeds of the Hermit Shale and Supai Group, units that make up red-brown ledges and shelves and that as they erode paint the great Redwall Limestone cliff to

match. The karst erosion surface at the top of the Redwall Limestone is rough and irregular, with solution cavities, broken rock, and red-tinged soil similar to limestone-derived soil of the tropics today.

Lees Ferry and Navajo Bridge. Lees Ferry is the launching site for Grand Canyon float trips. The river here is wide and slow moving. Cream-colored rock ledges near the river belong to the Kaibab Formation, very sandy here near the shore of the Permian sea in which it was deposited. Above Lees Ferry the lowest part of the Vermilion Cliffs is formed of Triassic strata: the dark red Moenkopi Formation at the base, and the variously colored Chinle Formation just above. Higher cliffs expose Jurassic sandstone with dune-type cross-bedding and blue-black desert varnish apparent on many rock faces.

Pima Point (South Rim). At the west end of South Rim Drive, Pima Point offers a good view westward into less accessible parts of Grand Canyon. Northwest across the canyon is Point Sublime, and to the northeast, about 12 kilometers (8 miles) away, Tiyo Point.

Rock units near and below Pima Point are virtually the same as those elsewhere in eastern Grand Canyon. The Kaibab Formation at the rim contains large Permian brachiopods and knobby masses of chert. Cross-bedding in the Coconino Sandstone shows up particularly well in cliffs just west of Pima Point.

Point Imperial (North Rim). Some 18 kilometers (11 miles) north of Cape Royal, Point Imperial looks down on wedges of Precambrian sedimentary rocks exposed along the East Kaibab Monocline. North of Point Imperial, Paleozoic formations above the Redwall Limestone have been eroded away from parts of the monocline.

Toroweap Point (North Rim). In western Grand Canyon the pattern of slopes and cliffs changes. The Esplanade, surfaced with soft red sandstone of the Supai Group, replaces the Tonto Platform as the widest shelf in the canyon. A thousand nearly vertical meters (3000 feet) below, the Colorado River flows in an inner canyon walled with Redwall Limestone. Cambrian and Precambrian rocks are deep beneath the river.

About a million years ago, great Niagaras of lava cascaded over the canyon's rim and into the river here. Coming from nearby Mount Trumbull, lava streams three times dammed the Colorado River, backing it up in long, narrow lakes. And three times the Colorado eventually cut through the lava dams. White splotches of lake-deposited clay still mark the canyon walls. A younger cinder cone, Vulcans Throne, is right at the brink of the canyon. Eruptions occurred down in the canyon as well as on the rim.

These volcanic features lie along the Toroweap Fault, which cuts across the canyon here. Toroweap Point is east of the fault, on its upward-moving side. Across the canyon the line of the fault and displacement of the layered rocks are quite apparent. Toroweap Fault extends northward beyond the Utah border, becoming the Sevier Fault.

Yavapai Point (South Rim). Here Grand Canyon is about 15 kilometers (9 miles) from rim to rim. From the museum balcony, Phantom Ranch can be seen in the green patch near the mouth of the long, straight gash of Bright Angel Canyon. With binoculars one can often pick out hikers and mule trains on slender threads of trails that cross the Tonto Platform 1200 meters (4000 feet) below.

Paleozoic sedimentary rocks of the upper canyon walls display their characteristic patterns, weathering into cliffs, ledges, shelves, and slopes, all the effects of differential erosion of resistant and less resistant rock layers. In general, limestone and sandstone form cliffs and ledges, siltstone and shale form slopes. Dark metamorphic rock—the Vishnu Schist—makes up the highest cliffs of the canyon, the walls of the Inner Gorge.

West of Bright Angel Creek, Precambrian sedimentary rocks appear beside the river. The eye-catching orange-red Hakatai Shale and the dark Cardenas Lavas are the most easily identified.

Exhibits here explain some of the canyon's other features.

OTHER READING

Anonymous, 1980. *Geologic Map of the Eastern Part of Grand Canyon National Park, Arizona.* Grand Canyon Natural History Association/Museum of Northern Arizona, Flagstaff, Arizona.

Beal, Merrill D., 1978. *Grand Canyon: The Story Behind the Scenery.* KC Publications, Inc., Las Vegas, Nevada.

Breed, William J., 1975. *Geologic Cross Sections of the Grand Canyon—San Francisco Peaks—Verde Valley Region, and of the Cedar Breaks—Zion—Grand Canyon Region.* Zion Natural History Association, Springdale, Utah.

Breed, William J.; Roat, Evelyn; and others, 1976. *Geology of the Grand Canyon.* Museum of Northern Arizona and Grand Canyon Natural History Association, Flagstaff, Arizona.

Butchart, Harvey, 1970. *Grand Canyon Treks.* La Siesta Press, Glendale, California.

Butchart, Harvey, 1975. *Grand Canyon Treks II.* La Siesta Press, Glendale, California.

Hamblin, W. Kenneth, and Rigby, J. Keith, 1968 (part 1) and 1969 (part 2). *Guidebook to the Colorado River.* Brigham Young University Geology Studies, Provo, Utah.

Loving, Nancy J., 1981. *Along the Rim.* Paragon Press, Salt Lake City, Utah.

Powell, John Wesley, 1895 (republished 1961). *The Exploration of the Colorado River and its Canyons.* Dover Publications, Inc., New York.

Rabbit, Mary C., 1978. *John Wesley Powell's Exploration of the Colorado River.* U.S. Geological Survey, Government Printing Office.

Mesa Verde National Park

Established: 1906
Size: 211 square kilometers (81 square miles)
Elevation: 1836 to 2531 meters (6025 to 8305 feet)
Address: Mesa Verde National Park, Colorado 81330

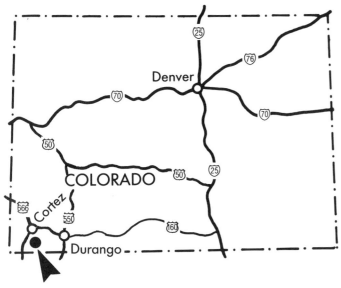

STAR FEATURES

• A high mesa whose walls portray the encroachment and departure of an oscillating Cretaceous sea.

• Close looks at erosional processes leading to formation of caves in horizontal strata.

• Views of the Four Corners country to the south, including the famous landmark Ship Rock, the San Juan Basin, and the broad furrow of the San Juan River.

• Archaeological remains, on the mesa surface and in large rock alcoves, left by Anasazi peoples who inhabited this region from 900 to 1250 A.D.

• Visitor centers, museum, introductory program, two loop drives, guided walks, self-guide trails, and shuttlebus trips to ruins. At most ruins visitors must be accompanied by park interpreters.

SETTING THE STAGE

Like Chaco Canyon, Mesa Verde National Park is primarily of archaeological interest—an unusual in-depth record of a prehistoric culture. However, the geologic story of the mesa is of interest as well.

A high, southward-sloping tableland that drops off precipitously in all directions, Mesa Verde is composed entirely of Cretaceous rocks. Its steep slopes reveal 600 meters (nearly 2000 feet) of gray siltstone: the Mancos Shale, deposited fairly far

from shore in a Cretaceous sea. This formation contains fossil marine shellfish as well as occasional bones and teeth of fish or of large marine reptiles known as plesiosaurs.

Weak and easily eroded, prone to landslides (to the detriment of the park entrance road), the Mancos Shale is capped with much harder rock—sandstones of the Mesaverde Group. This group, once considered a single formation, is now subdivided into three new formations, which we can look at in bottom-to-top order, the order in which they were deposited.

The lowest is the Point Lookout Sandstone, the rock that caps the north end of Mesa Verde, including the prominent north-jutting spine of Point Lookout. About 120 meters (400 feet) thick, this unit is a fine-grained, cross-bedded sandstone, largely a beach deposit. It contains fossils of both land plants and marine shellfish.

Above the Point Lookout Sandstone, the Menefee Formation is primarily nonmarine, alternating layers of shale, siltstone, and sandstone, with numerous thin bands of coal. It varies from 100 to 250 meters (340 to 800 feet) in thickness, and appears in canyons of the southern end of Mesa Verde, where it forms slopes below the mesa cap.

Capping the southern part of Mesa Verde and sheltering the cliff dwellings that have made this locality famous, the Cliff House Sandstone is massive, fine grained, and cross-bedded—again a beach or coastal deposit. It contains a few thin layers of shale and coal, as well as marine fossils such as cephalopods, clams, snails, and fish teeth and scales.

Younger Cretaceous formations have been stripped away by erosion. The three formations of the Mesaverde Group dip southward at a slightly steeper angle than the gentle slope of the mesa surface. Because of this, the upper two units of the Mesaverde Group—the Menefee Formation and Cliff House Sandstone—are eroded from the north end of the mesa. At the south end, the lowest unit—the Point Lookout Sandstone—is below the surface, even below the floors of the deep canyons there. The beveling of these formations is thought to have taken place before Mesa Verde was separated from the San Juan Mountains to the north, when mountain-derived streams carved a broad southern pediment.

In floor plan the mesa is something like a human hand, with deep south-draining canyons separating its long fingers. Many of the canyons are parallel, their position controlled by parallel north-south joints in the rock. Along the walls of nearly all the canyons, erosion has undermined the massive rimrock, causing large blocks to fall

Spruce Tree House was built in a deep recess in the Cliff House Sandstone. Springs that aided in creating the recess also furnished water for 12th and 13th century inhabitants.

Slope-forming Mancos Shale, deposited in a Cretaceous sea, undermines a cliff of Mesaverde Group sandstone that represents the near-shore deposits of a retreating sea. L. C. Huff photo, courtesy of U. S. Geological Survey.

Along the northwest edge of Mesa Verde, streams that once drained from the San Juan Mountains to the north have been "beheaded" and deprived of their water source as erosion of the Mancos Valley cut Mesa Verde off from the mountains.

away. Where surface water percolates downward through the porous sandstone, it meets the impervious shales and moves sideways until, at the canyon walls, it emerges as springs. And undermining is particularly effective near springs. There, conveniently for Anasazi villagers, large arcuate caves or alcoves had formed as arching sheets of sandstone fell away. Flat-floored alcoves protected by thick ledges of sandstone, provided with building blocks fallen from the ceiling, fed by springs of clear water—what better sites for villages?

Several dikes occur within the park; their igneous rock is basaltlike and contains visible biotite crystals in a fine grained, dark gray matrix.

Vital to Mesa Verde's prehistoric inhabitants, the surface of the mesa is covered with fine, even-grained soil. Where the soil came from is somewhat of an enigma. It is too fine grained to have been derived from the Cliff House Sandstone that caps the mesa. But it is similar to soils in southeastern Utah that are thought to have started out as wind-deposited Pleistocene silt, product of the Ice Ages.

Mesa Verde's cliff houses were discovered during the 1880s by members of a government-sponsored geological and geographical survey of western territories. Archaeological work here began in the 1890s, and intensified after the area became a park in 1906. Early in the 1930s many of the ruins were dated by analyzing the tree rings of logs used for ceiling beams. By 1935, 4000 sites of prehistoric habitation had been identified, some on the mesa surface, some in caves below the rim.

GEOLOGIC HISTORY

Mesozoic Era. Though the early geologic history of this area can be read in the San Juan Mountains, Mesa Verde's story begins with the advance of the Cretaceous sea, the last sea to cover the interior of the continent. Creeping over a land that for all of Triassic and Jurassic time had been above sea level, the sea came from the east and northeast. At first, as the land subsided, beach deposits accumulated—pure quartz sands that eventually became the Dakota Sandstone, a thin but widespread rock that now surfaces the area around the town of Cortez and also appears in hogbacks both east and west of the Rocky Mountains.

With further subsidence, the shoreline crept westward, and the sea in the Mesa Verde area deepened. In it, fine soft muds accumulated, muds that were to become the Mancos Shale. Millions of years in the making, this shale is nearly as widespread as the Dakota Sandstone, but a great deal thicker.

When at last the Mancos sea retreated, coarser sediments accumulated along its shores to form the Point Lookout Sandstone. Above this unit the rocks of the Menefee Formation were deposited by meandering rivers and streams and in shallow marshes and bays. A good deal of plant material also accumulated there, later to be compressed and altered into coal.

As the land level fluctuated once more, another thick layer of near-shore sand, the Cliff House Sandstone, was deposited, the uppermost and youngest layer of the Mesaverde Group. Here again, not all the material that went into this unit was pure sand. In quiet bays and estuaries fine silt and clay also accumulated, forming thin, shaly layers that eventually played a part in undermining the massive cliffs of Cliff House Sandstone, creating the great alcoves of Mesa Verde.

After accumulation of these sediments the sea deepened again, leaving behind the fine gray muds of the Lewis Shale and the sands of the Pictured Cliff Sandstone, with a total thickness of about 1650 meters (5400 feet). Then, late in Cretaceous time, as the Rocky Mountains began their upward push, the sea drained away one last time.

Cenozoic Era. The rise of the Rockies lasted well into Cenozoic time. In what is now southwestern Colorado, a broad dome, 100 miles across, rose as one of the ranges of the Rockies. During early Tertiary time erosion whittled at this uplift, carving down through domed layers of Mesozoic and Paleozoic rocks and into underlying Precambrian rocks. Along the southern side of the dome, where sedimentary rocks dip gently southward, both the Pictured Cliff Sandstone and the Lewis Shale were slowly eroded away.

Starting about 40 million years ago, volcanic activity completely changed the nature of the great dome. Eruptions lasted nearly 30 million years, with incredibly violent outpourings that destroyed whole stratovolcanoes as they collapsed into partly empty magma chambers, only to have new volcanoes build on the ruins of the old. Gradually the volcanic rocks built into the high, rugged range of the San Juan Mountains, northeast of Mesa Verde. In places, magma that did not break through the crust cooled and hardened into intrusive igneous rock.

Late in the history of this area, the entire region, from New Mexico to Montana, was lifted some 5000 feet, to its present elevations. As always, uplift increased erosion. With new vigor, streams draining the San Juan Mountains cut into surrounding areas, separating Mesa Verde and several similarly isolated mesas from the mountains. As the San Juan River deepened its valley, small tributaries draining the southern part of Mesa Verde eroded headward, creating the many narrow canyons that separate the southern fingers of the mesa. Aided by natural springs, the streams undermined the Cliff House Sandstone, creating the big alcoves that later became Anasazi building sites.

Erosion continues here today. Soft sediments give way in landslides along the margins of the mesa; large blocks of sandstone are undermined and tumble to the slopes below. Lichens establish themselves on rock surfaces, beginning the development of soil. Tree roots pry rocks apart. Rain and snowmelt initiate rivulets that course down steep slopes, carving rills that will turn into gullies that will turn into canyons. The tips of the mesa fingers will be cut off from the main mesa, surviving for a time as buttes and pinnacles, then disappearing altogether.

OTHER READING

Wenger, Gilbert R., 1980. *The Story of Mesa Verde National Park.* Visual Information Center, Inc., Denver, Colorado.

Yandell, Michael D., 1975. "Mesa Verde National Park." In *National Parkways Photographic and Comprehensive Guide to Mesa Verde and Rocky Mountain National Parks.* Worldwide Research and Publishing Co., Casper, Wyoming.

Natural Bridges National Monument

Established: 1908
Size: 20 square kilometers (12 square miles)
Elevation: 1980 meters (6500 feet) at visitor center
Address: c/o Canyonlands National Park, 446 Main Street, Moab, Utah 84532

Slender Owachomo Bridge formed at the junction of White and Tuwa canyons.

STAR FEATURES

• Three natural bridges, each showing a different stage in bridge development, carved in cross-bedded Permian sandstone.
• Numerous other erosional features, as well as examples of soil development in a semiarid climate.
• Triassic rocks mainly but not entirely eroded off the mesa surface.
• Visitor center, museum, introductory slide show, loop drive to the bridges.

SETTING THE STAGE

In White and Armstrong Canyons, which cut deeply through the horizontal surface of Cedar Mesa, three natural bridges have formed in rocks of the light-colored, cross-bedded Cedar Mesa Sandstone. Natural bridges, in contrast to arches, span the stream courses responsible for their formation. They can develop by several methods. Here, three aspects of the geologic setting have encouraged bridge formation:

• Rock units are horizontal.
• Streams that once wound across the mesa surface have retained their sinuous courses even as they cut deep into the sandstone. Entrenched meanders thus loop around narrow fins of rock extending from the mesa surface.
• The Cedar Mesa Sandstone is conducive to arch formation. It is composed of cross-bedded dune sand with fine, evenly sized grains cemented together with calcium carbonate. The cross-bedding is interrupted by horizontal layers of fine siltstone deposited in low areas between the former dunes.

* * * * *

In the semiarid climate of southern Utah, the streams in these canyons, though often dry, occasionally sweep forcibly against these jutting rock fins, grinding out the soft siltstone layers of the interdunes and pounding the sandstone with sand, pebbles, and boulders. Gradually the fins become thinner until finally they break through, leaving thick overhead arches—natural bridges.

Flowing water seeks the most direct path downslope. Once a bridge has formed, the stream abandons its meander loop and shortcuts under the bridge. As time goes on, it further undermines the rock and enlarges the opening below the bridge. With rain and snow, with day-to-night and season-to-season temperature changes, thin spalls of rock flake off from the undersurface of the span. Gradually the massive

Natural bridges span the streams that give them birth. Here, Sipapu Bridge arches across White Canyon.

bridge becomes thinner and thinner until only a narrow arch of rock frames the sky above the stream course.

The three large bridges featured in this national monument represent three stages in the life of such natural bridges. Kachina Bridge exemplifies an early stage. It is thick and bulky, with a relatively small passage below. Sipapu Bridge is thinner, more graceful. The stream is no longer eroding its abutments, but erosion of the bridge itself is continued by rain, wind, and frost. Owachomo Bridge is in a late stage of bridge development. Its span is slender and increasingly fragile. In the end, gravity will become too great a stress for this delicate span, and Owachomo Bridge will fall. As Sipapu and Kachina Bridges similarly become more and more delicate, other fins, pounded by flooding streams, will break through to create new bridges.

The rock in which the bridges took shape—the Permian Cedar Mesa Sandstone—shows many other interesting features, easy to see along the bridge trails. The steep dune-type cross-bedding is interrupted here and there by contorted wiggles that show where sand slumped or avalanched down lee faces of dunes. Even though the sandstone is of Permian age, at least 250 million years old, it is not very tightly cemented. You can rub off sand grains with your hand, and the simple patter of raindrops is enough to erode away a little of the loosened surface sand.

Younger sedimentary rocks have eroded off Cedar Mesa, though some of them are exposed on nearby buttes and mesas: dark red siltstone and mudstone of the Triassic Moenkopi Formation, just above it the white ledge of the Shinarump Conglomerate, and above that the purplish and greenish shales of the Chinle Formation—also Triassic. The mesa to the north is capped with a cliff of red-brown Wingate Sandstone darkened with desert varnish.

The drainage pattern here is controlled by two

Natural bridges develop where streams hammer at narrow fins of rock. At this entrenched meander or gooseneck, the stream may eventually carve another bridge. Lichens and desert varnish streak the barren rock.

sets of joints, one trending northwest-southeast, the other northeast-southwest. In many areas within the national monument, horizontal rock surfaces are marked with small potholes that contain little pools of rainwater. As these hollows develop, tiny organisms come to inhabit the pools, and their metabolic processes add acids that further dissolve the rock. As the pools dry up, the organisms curl up in cysts or lay drought-resistant egg clusters that will come to life again when their miniature worlds are once again moistened by rain.

Note the tapered black bands of lichens that mark seepage lines on many cliffs, both on Cedar Mesa Sandstone near the bridges and on nearby cliffs of Wingate Sandstone. Lichens are considered plant world pioneers in that they initiate processes that break rock down into soil. Shinier blue-black areas are desert varnish, thin deposits of iron and manganese. Smaller patches of desert varnish and more brilliantly colored lichens mark many other rock faces.

Puffy pink soils on the mesa are derived at least in part from the Moenkopi Formation. Gypsum, also from the Moenkopi Formation, is responsible for soil puffiness.

GEOLOGIC HISTORY

Paleozoic Era. There is no evidence in the national monument of the earliest history of this area, but most of its Paleozoic story is probably similar to that of the rest of the Colorado Plateau: advances and retreats of a shallow sea across the beveled surface of Precambrian igneous and metamorphic rocks. In Permian time, as the land rose above the sea, near-shore sand dunes swept across the region. Clusters of dunes were separated by flat interdune areas where silt and mud occasionally accumulated. Their product: the Cedar Mesa Sandstone and its related siltstone and mudstone interbeds.

Mesozoic Era. Early in Triassic time, an advancing sea beveled the top of the dunes and depos-

Whirled by streams, sand and larger rock fragments grind out potholes in rocky streambeds.

ited the mudstone, siltstone, and sandstone layers of the Moenkopi Formation. Here in southeastern Utah we are close to the ancient shoreline; the sediments are relatively thin and relatively coarse grained. Their source was in uplifts in Colorado and northern New Mexico. Raindrop marks and mud cracks show that parts of the formation accumulated above sea level, perhaps on tidal flats or along the margins of a large delta.

Later in Triassic time the Chinle Formation accumulated in near-shore lagoons. The climate may have become wetter. Quantities of volcanic ash frequently drifted over the area from volcanoes somewhere to the west, southwest, or northwest. South of here the great logs of the Petrified Forest were buried in muds of the Chinle Formation, and farther east and southeast the oldest dinosaurs left their skeletons in similar muds. Then desert conditions swept the region again, leaving behind the cross-bedded, dune-formed Wingate Sandstone.

There is little or no Jurassic or Cretaceous record in this area now, but we know from surrounding areas that a Cretaceous sea swept across it, depositing both near-shore and marine sediments. The shoreline was somewhere near the Natural Bridges area.

Cenozoic Era. As the Rocky Mountains rose late in Cretaceous and early in Tertiary time, this area received sediments washed from them. However, no Tertiary rocks remain within the monument, or even within sight of it.

By Quaternary time, streams draining the west slope of the Rockies wound across a wide, low plain, swinging in easy meanders, stripping away the soft Tertiary sediments. Strengthened by regional uplift about 10 million years ago, by rapid downcutting as the Colorado River deepened Grand Canyon, and by increased stream flow during the Pleistocene Epoch, still-winding streams incised their channels, cutting into flat-lying Mesozoic rock layers. As they cut "goosenecks" or entrenched meanders into harder Paleozoic strata—in this region the Cedar Mesa Sandstone—the cliffs of Mesozoic rock retreated, baring the harder Permian unit as the mesa surface.

As a final touch, the streams broke through fins of rock enclosed by the goosenecks, creating the natural bridges of this monument.

Prehistoric man—hunter-gatherers who left little sign of their passing—came into this region around 10,000 years ago. More complex cultures of the Anasazi peoples developed on Mesa Verde and at Hovenweep and Chaco Canyon to the east and south. Some of the Anasazi people built storage and sleeping rooms in alcoves below the rim of White Canyon, and left rock paintings near the natural bridges.

Navajo National Monument

Established: 1909
Size: 1.4 square kilometers (0.6 square mile)
Elevation: About 2221 meters (7286 feet) at visitor
center
Address: H.C. 71, Box 3, Tonalea, Arizona 86044

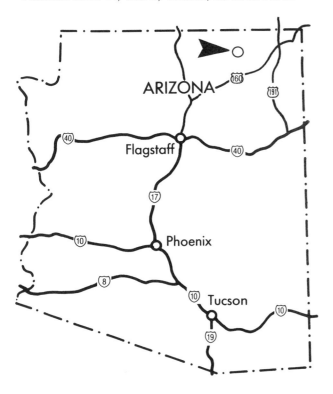

STAR FEATURES

• Prehistoric Indian villages—among the largest in America—protected from the ravages of time by arching overhangs of Navajo Sandstone.

• Smaller alcoves and other geologic features characteristic of this formation, visible near the entrance road and on the surface of the Shonto Plateau.

• Visitor center, museum, introductory slide show, trails, guided tours to the ruins (summer only). Visits to the ruins must be accompanied by Park Service personnel. At this writing, Inscription House is not open to the public.

SETTING THE STAGE

In dramatic Plateau country settings, the many-roomed "apartment houses" of Betatakin, Keet Seel, and Inscription House occupy unusually large, deep alcoves in massive, salmon-colored cliffs of Navajo Sandstone. The wide-sweeping cross-bedding and rounded, fine-grained sand of this formation indicate that it is a relic of former

sand dunes, part of an ancient Sahara that stretched from southern Nevada into northern Arizona, Utah, Colorado, and Wyoming.

The Navajo Sandstone is one of the major scenery-makers of the Colorado Plateau: It forms the great rock span of Rainbow Bridge, the towering cliffs of Zion Canyon, the bare-rock summit of Capitol Reef, the barren slopes that support the arches of Arches National Park. Here it bends up sharply in the Organ Rock Monocline near the monument entrance road, then levels out into the Shonto Plateau. The formation lends itself well to development of alcoves like those within this national monument, not just the large alcoves of the cliff dwellings but also the many small ones visible along trails to Betatakin and Keet Seel.

Triassic rocks below the massive sandstone—red siltstone and mudstone layers of the Kayenta Formation—also contribute to development of the big alcoves. They are relatively impervious to rainwater and snowmelt that seep downward from the surface through the porous Navajo Sandstone. Along the walls of deep canyons, where the contact between the two formations is exposed, the water comes to the surface in lines of springs. And where the Kayenta Formation is kept damp by these springs, it weakens and eventually washes away, undermining the massive sandstone above. Where it is undercut, the Navajo Sandstone breaks away in curving sheets and tumbles into the canyon below. Thus, the alcoves tend to mark the position of springs and seeps—no doubt an important factor in making these sites attractive to prehistoric peoples.

Smaller alcoves form in much the same way. Thin horizontal layers of reddish siltstone, thought to represent flat interdune areas where fine dust and silt collected, interrupt the smooth cross-bedding of the Navajo Sandstone. These thin siltstone layers also deflect the downward flow of moisture and initiate development of alcoves. In these small recesses, grains of sand and small rock slabs fall away, commonly leaving cross-bedding surfaces as the alcove ceilings. Animal tracks and droppings show that these alcoves, too, have been used as habitations.

Betatakin Canyon below the visitor center gradually erodes headward by the same process that creates the alcoves: gradual undermining of Navajo Sandstone. Springs that head the canyon, and the stream issuing from the springs, help to move loosened particles of rock material downstream. The stream may seem too small for such work, but its normal flow is occasionally supplemented by thunderstorm deluges.

There are several fine-grained, gray, freshwater

Weathering and erosion accent joints that score the surface of the Navajo Sandstone. The roots of pinyon and cactus hold enough soil for an island of vegetation on the otherwise barren rock.

Tree islands may contain remnants of soil that once covered the plateau surface. Climate changes and grazing (no longer permitted in the national monument) destroyed many of the original plants, whose roots held soil in place.

Betatakin (left) and Keet Seel (above) occupy large recesses in cliffs of Navajo Sandstone. The sloping ceilings are marked with layering of ancient dunes and streaks of black lichens. Keet Seel photo by John K. Loleit.

limestone layers within and at the top of the Navajo Sandstone here. They show us that shallow ponds developed in some interdune areas. Limestone layers along the Sandal and Betatakin trails are as much as 2 meters (6 feet) thick, and contain small nodules of black chert.

In places the sandstone cliffs are streaked with black, most of it due to the growth of lichens along seepage lines, where moisture is a little more plentiful than elsewhere. Don't confuse these dark, V-shaped, nonshiny bands with the shiny, blue-black desert varnish that coats some rock surfaces. Desert varnish accumulates particularly

along joints that developed during uplift of Shonto Mesa and of the entire Plateau region.

Barren slopes near the entrance road and on Shonto Mesa near the visitor center show that many joints, arranged in parallel sets, cut the Navajo Sandstone. One set runs almost north-south, another northeast-southwest. On the mesa surface these joints are deeply incised and serve as watercourses down which tiny rivulets flow during rainstorms. Small pools have developed along them in places, inhabited by tiny organisms that live out their lives in the short days or weeks after a rain, before the pools dry up. Acid by-products from the metabolism of these organisms help to dissolve the calcium carbonate that cements sand grains together, further enlarging the pools. Northeast-southwest joints controlled the position and direction of Betatakin Canyon and several other canyons in this area.

Bare rock slopes are rounded by the gradual peeling off of flakes of rock in a process known as exfoliation. Small patches of red, sandy soil develop in low spots or at the junctions of joints. Sage, grasses, pinyon trees, and junipers take possession of these patches; their roots penetrate the joints for nourishment. And while the roots and branches help to conserve the soil, they also break up new soil material.

GEOLOGIC HISTORY

Mesozoic Era. Since none of the early geologic history of this area is evident at Navajo National Monument (there is no reason to think it was different from that of the rest of the Colorado Plateau), we'll start our story in the Mesozoic Era.

During Triassic time this part of Arizona was a broad, gentle slope between the eroded remains of the ancestral Rocky Mountains and the shores of a western sea. The land was low and at times received fine sediments brought by rivers from the east and southeast. At least once during Triassic time it was inundated by the sea. Then, as

Large-scale cross-bedding formed as wind-blown sand accumulated on the downwind slopes of Triassic-Jurassic dunes.

the sea backed away, silt and mud that would eventually form the red rocks of the Kayenta Formation accumulated on the coastal plain.

This part of the continent then lay at about the same latitude as the Sahara and other great deserts of today—regions where peculiarities of atmospheric circulation lead to dryness. So with the retreat of the sea late in Triassic time, deserts spread along the coast and far inland from the coast, forming a sea of sand 400 kilometers (250 miles) across and 1000 kilometers (600 miles) from north to south. As sand dunes drifted across the desert region, fine silt and clay accumulated in flat interdune areas. Eventually the sand and interdune deposits became the Navajo Sandstone, here about 300 meters (1000 feet) thick, a formation that seems to span the Triassic-Jurassic boundary.

The region's later Jurassic history was one of changing environments and more incursions of the sea, with stream, lagoon, and marsh deposits as well as marine sediments. Some of the sediments carry significant amounts of volcanic ash.

In Cretaceous time a shallow sea invaded from the east; in it were deposited the rocks that now make up Black Mesa, the tree-covered tableland south of Shonto Mesa. Toward the end of Cretaceous time this sea drained away as the land rose again, signaling the development of the Rocky Mountains.

Cenozoic Era. If Cenozoic rocks were deposited here, they have been eroded away. Drainage in Eocene time was toward large interior lakes in central Utah. With uplift of the Colorado Plateau in Miocene time, with its segmentation into smaller individual plateaus, and with development of the Colorado River and the cutting of the Grand Canyon, drainage patterns changed and streams bit deeply into the Plateau surface. Canyon cutting may also have increased in Pleistocene time because of increased precipitation.

Prehistoric man found his way into this area about 10,000 years ago. Hunter-gatherers were followed by Anasazi peoples who built Betatakin, Keet Seel, and Inscription House. These ruins, dated by analysis of tree rings in their roof beams, were constructed around 1250 A.D., and abandoned, probably because of environmental and social problems such as prolonged drought or too frequent raids by their neighbors, only 50 years later. Their inhabitants may have migrated southeastward to pueblos along the Rio Grande, or southward to the pueblos of the Hopi Mesas. Despite the name of the national monument, they were not related to today's Navajo Indians.

Petrified Forest National Park

Established: 1906 as a national monument, 1962 as a national park
Size: 378 square kilometers (146 square miles)
Elevation: 1617 to 1900 meters (5305 to 6235 feet)
Address: Petrified Forest National Park, Arizona 86028

STAR FEATURES

• Numerous petrified tree trunks, their wood filled in and replaced, cell by cell, with jasper, agate, amethyst, and other colorful forms of silica.

• Other fossils, including fern, cycad, and ginkgo leaves, reptile and amphibian bones, and animal tracks. In 1985 the skeleton of a Triassic dinosaur was unearthed here.

• The Painted Desert, its subtle color patterns a legacy of stream and rivulet erosion in a many-hued palette of siltstone, mudstone, and volcanic ash.

• Visitor center, explanatory film, museum displays, a road guide to points of interest, nature trails, and talks by park naturalists.

See color pages for additional photographs.

SETTING THE STAGE

The Painted Desert of northern Arizona owes its existence not just to lack of rain, but also to the character of the Triassic rocks that form it. Layer upon layer of mudstone, siltstone, and claystone, with liberal additions of volcanic ash, surface this

Rafted by flooded streams, battered tree trunks slowly turned to stone as silica impregnated the wood. Desert pavement, with many pebble-sized fragments of petrified wood, discourages further wind erosion. National Park Service photo.

desert region. Poorly consolidated, these rocks swell when wet into tacky mud relatively impervious to additional water. Dry, the same surfaces are caked, cracked, and puffy, crunching underfoot and readily crumbling to dust. Unable to anchor themselves in the swelling, shrinking, rapidly eroding soil, few plants survive.

The volcanic ash in these Triassic sediments—the Chinle Formation, about 220 million years old—is altered now to clay minerals collectively called bentonite. One of the bentonite's constituents, montmorillonite, is the substance responsible for the swelling and shrinking. In arid and semiarid regions, bentonite encourages development of badlands, as it does here and in Badlands National Park, and parts of Yellowstone, Big Bend, Bryce, and several other national parks.

In the Painted Desert the badlands are delicately colored with small amounts of iron and manganese compounds and other minerals that occur as minor components of the volcanic ash and mudstone of the Chinle Formation. The soft rocks round into blue, blue-gray, purple, rusty red, white, and yellow mounds and ridges. Where they are capped and protected by hard layers of coarse sandstone and conglomerate or by large petrified logs, they stand as flat-topped, steep-sided ridges, buttes, and mesas, also important components of badland scenery. On such steep slopes, particularly just below overhanging ledges of harder rock, the soft layers are better exposed: They are truly rocklike, though they break easily into tiny chips.

In some parts of the national park the caprock is lava that came from two low-profile volcanic cones, Pilot Rock and Pintado Point. Other mesas here, as well as high ones east of the national park, are capped with Pliocene lake deposits.

The Painted Desert badlands extend far beyond the limits of Petrified Forest National Park, stretching west along the Little Colorado River to Grand Canyon. Near their western edge, where the Chinle Formation has been eroded away, deep red siltstone and sandstone of the underlying Moenkopi Formation give them added brilliance.

Petrified logs of this region are exceptionally abundant, exceptionally large, exceptionally colorful, and, unfortunately, exceptionally attractive to rock and mineral collectors. The national park was established with the express purpose of protecting them. Collecting *any* petrified wood—even small fragments—is prohibited. The specimens for sale in nearby curio shops come from the same geologic formation outside the park. (Indians now prohibit removal of fossil wood from Reservation lands.) The colorful badland scenery and petrified logs within the park are photogenic, and photographs make good souvenirs too!

The Petrified Forest is not a forest at all. Most logs are lying on their sides. Some seem battered, with limbs and roots broken off and bark scraped away. They are believed to have been downed trees that were rafted by floods or mudflows, rolled and buffeted, piled into logjams, and then rapidly covered with stream sediments and volcanic ash—much the same picture as that in Spirit Lake and along the Toutle River after the 1980 Mount St. Helens eruption.

Because volcanic ash is made of tiny fragments of unstable silica glass, groundwater seeping through the sediments soon becomes charged with dissolved silica. The silica tends to come out of solution when it contacts organic material such as old wood or animal bones. Little by little it has accumulated in pore spaces and cells within the trunks, bringing with it traces of iron, manganese, and other mineral substances that now add brilliant color to the wood.

Petrifaction is only one type of fossilization. Fern fronds and leaves from several types of trees occur here as imprints in layers of fine sediment. Fossil footprints of dinosaurs, other reptiles, and some large amphibians exist elsewhere in the Painted Desert but have not been found in the park. In some places, snail shells still retain their original calcium carbonate; others have left only molds, the original shell having been dissolved away. The skeleton of a Triassic dinosaur, like the trees silicified by percolating groundwater, was discovered here in 1984 and excavated in 1985. After it has been studied by paleontologists, it will be on exhibit in the park.

GEOLOGIC HISTORY

Mesozoic Era. The saga of the Petrified Forest and Painted Desert began in Triassic time, 225 to 220 million years ago, in the marshes and channels of a broad floodplain stretching between high mountains in southern or central Arizona and lower country to the north. In the mountains, luxuriant forests of pinelike trees related to araucarian "pines" of the present Southern Hemisphere sheltered small, graceful dinosaurs, crocodilelike phytosaurs, large, clumsy amphibians, dragonflies as big as swallows, and snails that crept among ferns of the forest floor. On the floodplain, tree-covered islands rose above shifting waterways and marshes, much as so-called "hammocks" rise above watery reaches of Everglades National Park today. In the distance, close to the mountains, wisps of fine volcanic ash drifted from clusters of volcanoes.

Now and then, over millions of years, sleeping volcanoes awoke with bursts of violence spawned far below in their magma chambers, and cloud after cloud of volcanic ash swept across this area. Forests smoked with heat, trees toppled, underbrush burned or was buried, animals perished. Perhaps snow on the mountains melted, or storm clouds, gathered by convection above the volcano's maw, deluged the land below. Floodwaters swept fallen trees far out onto the surrounding plain, buried their battered trunks, swept over them layers of ash and coarse sand and pebbles washed from the mountains. From time to time standing trees of the floodplain were inundated and preserved where they stood. More layers of silt, sand, and clay, more blankets of ash from more eruptions, accumulated above the entombed trunks.

The trees were buried so quickly that they did not decay. Instead, as silica-laden waters flowed through them, their pore spaces were little by little filled with silica. Where iron or manganese were present, some of the petrified wood took on subtle hues of agate and brilliant tones of jasper —white and yellow, red and black. In hollows within the trunks, quartz crystals had room to grow, some glassy and colorless, others tinted the pale lavender of amethyst.

Ultimately the mountains to the south wore away. In Cretaceous time, seas advanced again, coming not from the west as in Paleozoic time, but from the east. New deposits were laid down, thick shale and sandstone layers bearing fossils of marine creatures rather than of land animals.

About 70 million years ago, as the Cretaceous Period and the Mesozoic Era drew to their close, this region began to rise once more and the seas drained away for the last time. As the earth moved, the stresses and strains of upheaval cracked many of the buried, petrified trunks and segmented them like cordwood.

Cenozoic Era. For much of the time since the close of the Cretaceous Period, erosion has been the dominant force here. Originally near sea level and later lifted to its present elevation, the region has been at times grassland, at times marshy floodplain, at times covered with lava and new falls of volcanic ash. Gradually the Cretaceous marine sediments were washed away. Then for millions of years the region held a great lake, as large as Lake Erie today. In its waters, lake sediments accumulated, forming the Bidahochi Formation, now visible at the park's northeast

boundary. Then erosion gained the upper hand again, removing much of the lake sediment and lava and volcanic ash, and baring at last the old Triassic floodplain deposits: sandstone and mudstone and conglomerate, colorful bentonite, fossil reptiles and amphibians, insect impressions and leafprints, and the petrified logs of this national park.

BEHIND THE SCENES

Viewpoints and loops are described in north-to-south order, as they appear on entering the park from the north.

Painted Desert Loop. Observant eyes will see many geologic features here: badlands eroded in soft, bentonite-rich rock layers; subtle colors given to the Painted Desert by the minerals of nature's palette; ridges and mesas protected by hard sandstone layers and thin, relatively recent lava flows; low-profile volcanoes from which some of these lava flows came; undermined lava blocks breaking away from the mesas; sharp white veins of gypsum; deposits of white caliche.

Some of the young volcanic rocks along the mesa edge and near the picnic area are simple lava flows with many gas bubble holes or vesicles. Others include large blobs of basalt thrown forcibly from volcanic vents, rounded and football-shaped from their brief flight through the air.

Geologists believe that the gypsum layers formed here in shallow, evaporating lakes and ponds of the Triassic floodplain. Gypsum is a common constituent of many rock layers in the Southwest and dissolves fairly easily, so the streams feeding these ponds may have contained large amounts of the mineral. Gypsum in veins is a secondary deposit, leached by groundwater from gypsum layers within the rock and redeposited in slender cracks. The gypsum occurs as a transparent, silky-surfaced mineral called selenite. Salt, which also may have been deposited in the evaporating lakes, dissolves readily; if it formed here in Triassic time it has been removed by groundwater.

Newspaper Rock. This site holds interest for geologists as well as archaeologists because the petroglyphs were created by patiently pecking through the dark desert varnish of the rock face. Desert varnish is a shiny brown or blue-black coating of manganese and iron minerals derived from dust or from the rock itself. As rain or other moisture sinks into the rock, it dissolves and carries along tiny amounts of these minerals. Later, as the moisture evaporates from the sun-warmed

Left *and* below, *scattered logs, naturally broken into cordwood lengths, lie in deeply eroded ravines completely lacking in plant growth. Ray Strauss photos.*

Painted hills of Chinle Formation, rich in clays formed from volcanic ash, are too unstable for plant growth. The soft clays and fossil "forests" may be the result of explosive volcanic eruptions. National Park Service photo.

rock, it leaves its mineral burden, however small, at the surface. As you can see here, where petroglyphs were pecked through it hundreds of years ago, new desert varnish has not yet had time to form.

Blue Mesa Loop. The approach from the main park road to Blue Mesa heads almost straight toward a small, low volcanic cone, one that emitted very fluid basalt lava—hence the low profile of the cone.

A layer of cross-bedded sandstone, one of the coarser, more resistant layers of the Chinle Formation, caps Blue Mesa. Short, straight crossbedding in this layer is of the type formed by running water, so we know that sand was deposited in a stream channel or along the edge of a small delta, with layers of sand building downstream.

East and south of the south parking area, in a maze of little canyons, petrified logs perch on narrow ridges, for a time sheltering the soft rock that makes up their pedestals. When the pedestals finally wash out from under them, the logs will roll or tumble downslope. This "forest," with its large number of tree trunks, is thought to have been a log jam in the channel of a flooding Triassic stream. The blue, white, lavender, and pale green of eroded slopes remind us that this region is part of the Painted Desert. These colors form where oxygen is in short supply in the original sediment, as it often is in ash falls or in floodwaters containing abundant decaying plant and animal matter.

The view north from Blue Mesa reveals a different scene, with tones of brown and rust predominating, and with the edge of the mesa reaching north like the paws of a giant sphinx. Single lay-

ers in the Chinle Formation are not extensive, and are in many places replaced horizontally by strata of different color, grain size, or bentonite content.

The loop trail follows the summit of a ridge and then drops downward close to these rocks. Notice the desert pavement along the ridge, with small round pebbles of hard igneous and metamorphic rock or gray chert. The pebbles were washed here in Triassic time from highlands in central Arizona, about 100 kilometers (60 miles) south and southwest of here. The chert comes from Permian rock—the Kaibab Formation, now absent in central Arizona. Igneous and metamorphic pebbles are mostly Precambrian, and provide important evidence of the existence of high mountains there in Triassic time. Studies show that the pebbles become larger southward, toward their source.

From the trail, watch for large slump blocks that are sliding down and away from the edge of the mesa. Some tilt back toward the mesa, showing that they moved on curving slide planes.

Agate Bridge. At this site a petrified log spans a small, usually dry watercourse. The ends of the log rest in the sandstone in which it was originally entombed. The concrete support was added in 1917, when it was feared the heavy stone log might collapse. Several other log spans occur in the park.

Jasper Forest. A thin layer of pebbles covers much of this area. As at Blue Mesa, many of the pebbles are of recognizable Precambrian metamorphic and igneous rocks, or of chert from the Kaibab Limestone. Washed in Triassic time from mountains in central Arizona and deposited as conglomerate in the Chinle Formation, they have now been recycled into desert pavement. Other pebbles are small angular fragments of petrified wood. The pebbly pavement is in places cemented by hard white caliche, the "hardpan" of desert regions, deposited near the surface as lime-laden groundwater evaporated. Cemented or not, the desert pavement protects underlying clay and siltstone from erosion. It also helps rainwater sink into the ground rather than flow off across the surface, and forms an armor that lessens evaporation.

Crystal Forest. The name of this "forest" stems from crystals of quartz and amethyst in cavities of the petrified logs. Unfortunately, most of these crystals were extracted before the area came under national park protection. Mineral collectors went so far as to blast many trunks open to get at the clear, well-formed crystals.

Flattops. The pebbly surface here does not everywhere coincide with a hard caprock. In Pliocene time a stream's broad floodplain covered this part of the national park, forming this surface. The stream and its tributaries later cut down into the floodplain, leaving only the present flat-topped mesa as a remnant of the old floodplain. The increased erosion may have been due to rapid downcutting in Grand Canyon and in the gorge of the Little Colorado River, as well as to increased precipitation during Pleistocene rainy cycles.

Rainbow Forest and Museum. Exhibits here display fossil leaves and reptile bones found in the park, as well as polished petrified wood showing its vivid colors and the microscopic details of the original wood.

The museum is near an unusually dense concentration of large petrified logs, many of them brilliantly colored with red and yellow jasper. Some show remnants of their original bark, often a clue, as the exhibits show, to the kinds of trees involved.

OTHER READING

Ash, Sidney R., 1986. *Petrified Forest, the Story behind the Scenery.* Petrified Forest Museum Association, Holbrook, Arizona, and KC Publications, Inc., Las Vegas, Nevada.

Bezy, John V., and Trevena, Arthur S., 1975. *Guide to Twenty Geological Features at Petrified Forest National Park.* Petrified Forest Museum Association, Holbrook, Arizona.

Colbert, E.H., and Johnson, R.R., 1985. *The Petrified Forest through the Ages.* Bulletin series 54, Museum of Northern Arizona Press, Flagstaff.

Rainbow Bridge National Monument

Established: 1910
Size: 0.64 square kilometer (0.25 square mile)
Elevation: Lake Powell high-water level, 1128 meters (3700 feet)
Address: c/o Glen Canyon National Recreation Area, P.O. Box 1507, Page, Arizona 86040

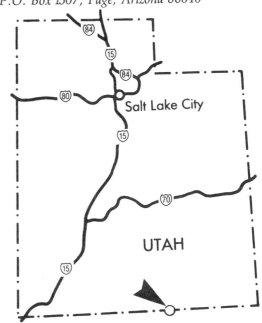

STAR FEATURES

• The world's largest known natural bridge, carved by stream erosion amid a landscape of barren cliffs and canyons.

• Sedimentary and erosional characteristics of the dune-formed Navajo Sandstone, as well as those of rock formations above and below it.

• Views of Navajo Mountain, a large laccolith.

• Visitor center (at Glen Canyon Dam) with exhibits and introductory film; boat dock and short trail to Rainbow Bridge.

SETTING THE STAGE

More accessible now than it was before Lake Powell filled, Rainbow Bridge, deep in a slickrock wilderness of rugged canyons, arches across the channel of Bridge Creek, one of several streams that drain the north slope of Navajo Mountain. The bridge, whose graceful 85-meter (278-foot) span soars 94 meters (309 feet)—the height of a 30-story building—above the creek, is the largest rock span known.

Formed of salmon-colored Navajo Sandstone, a rock unit responsible for much of the scenic grandeur of the Plateau country, the bridge owes its shape to sedimentary and erosional characteristics of this sandstone. Long, sweeping cross-bedding, coupled with uniformly fine, rounded, frosted sand grains, show that the formation is wind deposited—the product of desert sand dunes. Sand grains are cemented together with calcium carbonate to form a strong sedimentary layer capable of standing in high vertical cliffs. En route to Rainbow Bridge by boat or by trail it is easy to see such cliffs, marked here and there with large, arching overhangs.

Above the Navajo Sandstone are thinly layered shale and siltstone of the Carmel Formation. This unit, best seen from Lake Powell and in road cuts between Glen Canyon Dam and Wahweap, is weak and easily eroded, wearing back into broad benches.

Above it, the Entrada Sandstone caps many of the buttes and promontories that jut into Lake Powell. Like the Navajo Sandstone, the Entrada is an eolian (wind-formed) rock unit, complete with large-scale diagonal cross-bedding and fine, well-rounded sand grains.

Below Rainbow Bridge, and contributing to its genesis, is the Kayenta Formation, a series of red sandstone and siltstone ledges that bear the features of river floodplain or delta deposits. In many places this rock is covered by lake waters; at Rainbow Bridge it is above water level and forms the horizontal ledges that support the bridge. Since its sandstones are hard and resistant, the Kayenta Formation as a whole erodes less easily than the overlying Navajo Sandstone. It is also less porous than the Navajo Sandstone, and deflects rainwater and snowmelt percolating down through the sandstone, causing springs and seeps along the contact between the two formations.

South of Rainbow Bridge, visible up Bridge Canyon, Navajo Mountain rises to 3166 meters (10,388 feet) above sea level. It, too, probably

Streaked and stained with desert varnish, the sculptured arch of Rainbow Bridge rests on narrow ledges of resistant sandstone. Both bridge and backdrop are carved in Navajo Sandstone. Ray Strauss photo.

A line of seeps, where mosses and lichens add black accents to the tawny rock, weakens a Navajo Sandstone cliff. A curving fracture (arrows) is developing above this weakened zone. The rock below it will eventually drop away, leaving an arching recess.

played a role in creation of the natural bridge. Navajo Mountain is a laccolith, a granite intrusion that about 65 million years ago pushed between layers of sedimentary rock, never reaching the surface but doming the rock units that lay above it. Strata that once covered Navajo Mountain like draped blankets have now eroded from its summit, but their eroded edges still encircle its base.

How and why did Rainbow Bridge form? A single, often overlooked feature near the bridge gives us a good clue. Behind the east buttress of the bridge, well above the present stream channel, is part of an abandoned older channel that tells us that Bridge Creek once took a sharp loop here, a "gooseneck" meander loop inherited from a slow, sluggish course across a gently sloping surface. As the stream, invigorated by plateau uplift and the rising mass of Navajo Mountain, cut ever deeper, it retained its gooseneck pattern, entrenching its meander loop deeply into the rock.

As it rushed through this twisting gorge, the stream pounded sand, pebbles, and boulders against the rock spur circumscribed by the entrenched loop. At some stage, as the spur be-

came thinner and thinner, the stream broke through it, abandoned its entrenched gooseneck, and took the more direct route downstream.

At first the bridge was probably thick-bodied and heavy, with only a small opening. But, helped by its steepened gradient, the stream continued to cut downward through the Navajo Sandstone. The resistant Kayenta Formation, below the Navajo, retarded downward cutting for a time, and the stream instead widened the passage below the bridge. Springs along the contact between the formations probably aided in the erosional process by keeping the rocks well saturated and subject to freeze-and-thaw weathering; they may also have helped to remove some of the cementing material that held the sandstone together.

Perhaps it was at this stage that large blocks of Navajo Sandstone fell away in the archlike pattern common in this area. With its opening thus enlarged, and with further deepening of the stream channel, bridge surfaces were further refined. Thin, curving sheets of rock spalled off them, gradually smoothing the great span—a process that will continue until the bridge becomes so narrow it can no longer support its own weight.

Thin-bedded, slabby, and resistant, the Kayenta Formation forms a precarious base for Rainbow Bridge. Ray Strauss photo.

GEOLOGIC HISTORY

Since rocks of the Precambrian and Paleozoic Eras, deeply buried here, play little or no part in the origin of Rainbow Bridge, we will start our story in Mesozoic time.

Mesozoic Era. Three rock units are exposed at this national monument: the Triassic Kayenta Formation, the Triassic-Jurassic Navajo Sandstone, and the Jurassic Carmel Formation, the last high up and out of sight from the canyon floor. These three units tell us of the changes from floodplain or delta to arid desert and back to river floodplain. Following their formation, thousands of feet of Cretaceous sedimentary deposits accumulated, some on land, some during pulselike advances of an eastern sea.

Cenozoic Era. More thousands of feet of sedimentary rocks may have accumulated here in Tertiary time: thick sandstones and siltstones made of rock material brought by streams draining the newborn Rocky Mountains. As their gradients steepened, tributary streams that had meandered sluggishly across a low-elevation plain became vigorous enough to strip away whatever younger sediments may have existed, and to incise their earlier gooseneck meanders into the underlying Mesozoic rocks. Stream vigor in the Rainbow Bridge area was further enhanced by upward doming of the Navajo Mountain laccolith about 65 million years ago, and by later downward erosion by the Colorado River.

As this great river and its small tributaries cut deep into Jurassic and then Triassic rocks, many narrow, winding canyons formed, including mystically beautiful Glen Canyon on the Colorado River, now hidden beneath Lake Powell's waters. Twisting salmon-colored cliffs and slopes of barren rock, etched by wind and rain, displayed the diagonal cross-bedding of Jurassic dunes and the brown and purple streaks of desert varnish. After the rainy cycles of Pleistocene time, this area once more became a desert, with sand from deserts of the past becoming ornaments in a desert of the present.

OTHER READING

Everhart, Ronald E., 1983. *Glen Canyon—Lake Powell: the Story Behind the Scenery*. KC Publications, Las Vegas, Nevada.

Sunset Crater National Monument

Established: 1930
Size: 12 square kilometers (about 4.8 square miles)
Elevation: 2120 meters (6958 feet) at visitor center
Address: Route 3, Box 149, Flagstaff, Arizona 86004

STAR FEATURES

• A classic cinder cone, geologically young and of near-perfect symmetry, with smooth gray slopes tinted at the rim in sunset colors. Jagged lava flows nearby are still fresh looking and almost unvegetated.

• Spatter cones, squeeze-ups, lava bubbles and tunnels, and other detailed volcanic features.

• Views of other cones of the San Francisco volcanic field, with San Francisco Mountain as the reigning monarch. All the features of this field developed during the last 6 million years.

• Interpretive film and exhibits, self-guided nature walks, evening programs.

SETTING THE STAGE

Sunset Crater greets visitors with a tinted summit due not to filtered evening light (though a colorful sunset heightens the effect) but to oxidation and fumarole (steam vent) deposits at its rim. The cone of dark gray basalt cinders is 300 meters (1000 feet) high. It is symmetrical in shape except for a lopsided rim caused by the leeward drift of cinders and ash before prevailing southwest winds. A blanket of cinders jettisoned by Sunset Crater extends far east, north, and south, covering an area of almost 2000 square kilometers (800 square miles).

A classic cinder cone, Sunset Crater erupted in 1064–1065 A.D. *Its colorful rim hides a deep central crater. Tad Nichols photo.*

The basalt cinders have not weathered significantly in this cool, dry climate; they are still crisp and crunchy underfoot. Nor is the crater's unprotected surface grooved by gullies, as you might expect, for it is so porous that rain and melting snow sink right in.

Two rough, black lava flows lie at the foot of the crater. The western or Bonito flow, the one most visitors see, fills the basin between older cinder cones. The Kana-a flow on the east side of the crater is long and narrow because it flowed down a pre-existing stream channel. Where it is exposed to view, the lava looks as though it oozed only yesterday from fissures at the base of the cone. But look again—the lava surfaces are spotted with lichens, pioneers of the plant world. With their help, and with increasing assistance from other plants, the rock will eventually change to soil. Shrubs and trees are already taking root in pockets of cinders and windblown pine needles. Each new tree or shrub or clump of grass will add to the growing supply of humus when it drops its leaves or needles, or dries in autumn.

In places the lava contains pieces of white or light gray Paleozoic limestone broken from the walls of the volcano's conduit as magma moved upward. Blocks of intrusive rock have been brought up, too, from intruded masses related to much older, subsurface igneous activity.

Flow patterns are well developed on the Bonito lava flow. Look for ropy pahoehoe lava; for rough, hard-to-walk-on aa lava; and for patterns of concentric pressure ridges curving across lava tongues. Both flows and cinders are basalt, and came from the Earth's mantle, rising along a fissure. Because it cooled fairly rapidly the rock is so fine grained you can't see individual crystals except for scattered little crystals of green olivine and white or transparent feldspar.

On the lava flows, squeeze-ups are a late development. Their parallel grooves and scratches record the manner in which they pushed up through jagged breaks in the hardening skin of the lava flows. Lava caves and tunnels developed where fluid lava flowed out from beneath its own hardening crust. Blisters and bubbles on the lava surface, some of them several meters across, formed as hot gases pushed upward against a still-flexible skin of cooling lava.

Trending northwest and southeast from Sunset Crater, a line of spatter cones defines the fissure from which the main crater's cinders and lava were emitted. Red-tinted spatter from these cones clearly overlies the black cinders of the parent crater, so we know that they represent a fairly late stage in the eruption story.

O'Leary Peak, north of Sunset Crater, is another expression of volcanism, a lava dome squeezed upward about 200,000 years ago as a doughlike mass of very thick, viscous lava. Considerably lighter in color than Sunset's dark flows, its lava is more silicic than that of Sunset Crater; in composition it resembles lavas of the San Francisco Peaks. The view from the summit takes in Sunset Crater, the Bonito flow, many other cinder cones, and the eroded stratovolcano of San Francisco Mountain itself. Eastward, O'Leary Peak overlooks the eastern edge of the Coconino Plateau, where Paleozoic sedimentary rocks bow downward and plunge beneath younger strata of the Painted Desert. (Wupatki National Monument lies on this great flexure; the story of its prehistoric villages ties in with the story of Sunset Crater's eruption.)

GEOLOGIC HISTORY

Precambrian, Paleozoic, and Mesozoic Eras. The Precambrian and Paleozoic history of this region is essentially identical to that of Grand Canyon, known from unequalled exposures there. Here in the San Francisco volcanic field these older rocks are concealed by a veneer of volcanic rocks. Dark red Triassic sandstone and mudstone originally covered this area too, as did several thousand meters of Jurassic and Cretaceous sandstone and shale, relics of ancient deserts and floodplains and one last incursion of the sea.

Cenozoic Era. With uplift of the Colorado Plateau in early Tertiary time, erosion became the main process shaping most of the landscape. Particularly attacking the Kaibab Uplift, one of the highest plateau segments, Cenozoic erosion stripped off layer after layer of Mesozoic sandstone and shale, laying bare the hard Paleozoic limestone below—a process still going on in the Painted Desert to the east and north.

But about 6 million years ago, during the Pliocene Epoch, volcanic rumblings warned that other landscape-formers were gathering strength far below the plateau surface. The first volcanic outpourings were of very fluid basalt that spread like a thin veneer over the southern part of the limestone-capped Kaibab Uplift. Then, about 1.8 million years ago, there followed a series of eruptions of thicker, more silicic lava, fairly light in color because of its high quartz and feldspar content. Lava flows and volcanic ash from innumerable eruptions piled up gradually into the stratovolcano of San Francisco Mountain. Lava domes and in some places laccoliths pushed upward, too. Finally, late in the history of the vol-

As volcanic pressures build up under a lava crust, pasty lava pushes through cracks to form squeeze-ups. Tad Nichols photo.

San Francisco Mountain's Inner Basin, roughed out by explosion or collapse, was smoothed by Pleistocene glacial erosion. Sunset Crater cinders blanket the foreground. Tad Nichols photo.

Spatter cones build up as fluid lava splashes to the surface. Tad Nichols photo.

canic field, long after the San Francisco volcano had been reshaped by explosion, collapse, and Ice Age glaciers, some 400 cinder cones erupted around its base.

Sunset Crater is the youngest of these cinder cones, the most recent expression of volcanic activity in the southwestern United States. Its cindery slopes and jet black flows are less than a thousand years old. We can readily reconstruct its birth and growth because Sunset Crater closely resembles a modern volcano, Paricutin, which erupted in west central Mexico between 1943 and 1954. Paricutin was born before the astonished eyes of a Mexican farmer as a tiny fissure in his cornfield began to emit strange rumbles and belch thin wisps of smoke. Becoming bolder, the new vent tossed out a few bits of spongy pumice. Then more, and more, until a cone began to form. Soon cinder and ash and spinning football-sized bombs of lava or lava-coated rock rained down mercilessly on cornfield, house, and surrounding country, killing all vegetation and driving local inhabitants from their homes. The new volcano of course attracted geologists and photographers, who recorded its growth in lasting detail. As frothy magma rose and was blown from the summit in a long succession of explosions, less gaseous magma began to rise in the conduit and eventually broke out through the base of the cone and streamed over the ash-covered countryside, engulfing finally the entire village of Paricutin.

Sunset Crater, too, was witnessed by man. It, too, deluged cornfields, as well as plots of beans and squash, staple crops of early Sinagua people who made their homes in this region. Several prehistoric pit houses have been unearthed by archaeologists, their roofs collapsed by the weight of the cinders, their owners fled from trembling earth and fiery rain. Tree-ring studies of charred roof timbers from these dwellings, as well as radiometric and paleomagnetic dating of the volcanic rock, tell us the eruption began after summer tree growth of 1064 A.D. and before summer growth of 1065—shortly before William the Conqueror set foot on English shores. This eruption may have lasted only a few months or it may have been spread out as a series of eruptions over several decades.

Like Paricutin, Sunset Crater lies in a setting studded with older cinder cones. As at Paricutin, lava flows oozed from its base. Here the Kana-a flow, northeast of the crater, developed at about the same time as the main cinder eruption in 1064/1065. The Bonito flow came later, and is variously dated as 1180 or 1220 A.D. Smaller eruptions later flung cinders onto the surfaces of these flows. Spatter cones formed as the cinder activity subsided. Near Sunset's crest, volcanic gases escaped from fumaroles for some time after other activity ceased, perhaps as late as 1220 or 1250, depositing gypsum and sulfur and reddening the rim with iron oxides, the last gaseous gasps of the dying volcano.

The culmination of Sunset's story involves displaced persons: the Sinagua people who fled its angry fountains. They and their descendants returned to find that volcanic ash and cinder, now cooled and no longer terrifying, have a mulching effect that curbs evaporation and conserves precious moisture, thereby adding to the productivity of fields and garden plots. In an early population explosion, they settled here anew and in ever greater numbers, building clustered, multi-storied apartment houses among the rocky fringes of the cinder fields, apartment houses such as those at Wupatki National Monument.

There is no reason to believe that the eruption of Sunset Crater, the youngest of about 400 cinder cones in this area, brings volcanism to a close here. Dating of other cinder cones in the San Francisco volcanic field shows that they become progressively younger toward the northeast, as if the Earth's crust had drifted slowly southwest over a stationary "hot spot" in the mantle. So the next eruption will probably be in the area northeast of Sunset Crater. When it comes, it will probably give warning in the form of earthquakes, steam vents, and perhaps swelling and cracking of the surface.

OTHER READING

Breed, William J., 1976. "Molten Rock and Trembling Earth." *Plateau*, vol. 49, no. 2.

Breed, William J., 1978. "The Mountains of Fire." *Arizona Highways*, vol. 54, no. 7, p. 24.

Lowe, Sam, 1978. "The Day the Earth Caught Fire." *Arizona Highways*, vol. 54, no. 7, pp. 2-5.

Schroeder, Albert H., 1977. *Of Men and Volcanoes—the Sinagua of Northern Arizona*. Southwest Parks and Monuments Association, Globe, Arizona.

Walnut Canyon National Monument

Established: 1915
Size: 9 square kilometers (3.5 square miles)
Elevation: 2072 meters (6800 feet) at visitor center
Address: Walnut Canyon Road, Flagstaff, Arizona 86004

STAR FEATURES

• Cliff dwellings built in recesses in Permian limestone layers similar to those in the upper walls of Grand Canyon.

• A deeply entrenched loop of Walnut Creek.

• A close look at the transition between two formations, and at weathering features in limestone and dolomite.

• Visitor center, slide show, illustrated trail guides, guided walks. Exhibits include a display explaining the science of dendrochronology or tree-ring dating.

SETTING THE STAGE

One of many small gorges that lace the Coconino Plateau, Walnut Canyon provides an interesting view of geologic processes of Permian time, as well as of the more recent past. Around 800 years ago, at the time when Sinagua Indians built homes in cliff recesses in the canyon, Walnut Creek flowed more often than it does now, and may have retained water in perennial pools even when the stream was dry. Now most of its

Small cliff dwellings peer from caves beneath massive ledges of silty and sandy Kaibab Limestone. The limestone here was deposited nearer to shore than equivalent strata in Grand Canyon.

flow is impounded in Lake Mary as part of Flagstaff's water supply.

Immediately below the visitor center the stream describes a tight loop that may originally have been a meander bend across a relatively even, low-gradient surface, or that may have originated in a twisting solution cavern in the limestone, a cavern that later collapsed. In either case, as the gradient was steepened the stream retained its looplike course but cut deeper and deeper—over 100 meters (350 feet) in all—into surrounding rock layers.

From the visitor center the view looks out on the incised stream loop and on the "island" it circumscribes. Beyond is the level surface of the Coconino Plateau, capped with resistant limestone of the Kaibab Formation, the same rock that caps the rims of Grand Canyon.

The Kaibab Limestone in the upper canyon walls, exposed as a series of ledges and slopes, is made up of thick, fairly massive layers of sandy or silty limestone and dolomite separated by thinner layers of limy, shaly siltstone and sandstone. (Dolomite differs from limestone in containing magnesium carbonate as well as calcium carbonate. When it weathers, it forms a rough, rather prickly surface.) Many of the layers bear marine fossils: clams, snails, bryozoans, and brachiopods. The nature of the limestone and dolomite, their fossils, and the amounts of silt and sand within and between them suggest a near-shore

but nevertheless marine environment.

The lower canyon walls are formed of fine white cross-bedded Coconino Sandstone, gray with lichens. This rock unit contains no fossils at Walnut Canyon, but displays the patterns of crossbedding characteristic of beach- and dune-formed sandstone. Like the Kaibab Formation, this part of the Coconino Sandstone reflects a near-shore environment; unlike the Kaibab, the sandstone formed on the shore rather than just offshore.

Many parallel vertical joints cut through the canyon walls, particularly in the Coconino Sandstone. To some extent they control the drainage here, partly guiding, for example, Walnut Canyon's sharp gooseneck bend. Steep ravines on the canyon walls are erosion-enlarged joints. Several faults cut across the canyon, with just a few centimeters of displacement. A fault near Site 6 on the trail to the ruins has a displacement of about 3 meters (10 feet).

The trail drops down across ledges of Kaibab Formation limestone and dolomite—some of it sparkling with grains of fine quartz sand, some patched or completely coated with lichens. Fossil shellfish, particularly hollow molds left where shells were dissolved away by groundwater, can be seen in several of the ledges. Such fossils are studied by filling the little hollows with latex, letting it dry, and then snapping the latex out of the hollows, giving perfect casts of the original shell. Salmon-colored geodes, dark brown limonite con-

Cross-bedding in the Coconino Sandstone reflects its dune origin. Vertical joints are strongly developed here and to some extent guided development of Walnut Canyon.

cretions, and nodules of hard white chert protrude from some limestone surfaces. The chert formed around sponges whose skeletons, made of tiny needles of silica, have left a fine tracery visible when the nodules are broken open.

Much of the trail around the "island" is just above the boundary between the Kaibab Formation and the Coconino Sandstone. There are unusually good exposures of the Coconino, with its prominent cross-bedding, across the canyon.

GEOLOGIC HISTORY

Paleozoic Era. Most of the Paleozoic formations of Grand Canyon's walls underlie Walnut Canyon. But the smaller, shallower canyon cuts through only the uppermost layers, so we'll begin our story fairly late in Permian time, when shallow seas twice crept in from the west. The two advances and retreats of the sea are well documented in Grand Canyon's upper walls, where each involves a sequence of shoreline deposits of the advancing sea, marine limestones containing abundant fossils, and then shoreline deposits again. But at Walnut Canyon, near the farthest eastward extent of the Permian sea, the first of these marine cycles is absent, replaced by the Coconino Sandstone, which contains no true marine sediments. The second cycle is represented by the silty, sandy limestone and dolomite of the Kaibab Formation, all of it deposited fairly near the shore.

Mesozoic and Cenozoic Eras. At the end of Permian time the sea retreated. Colorful Mesozoic strata—the Moenkopi, Chinle, and younger formations of the Painted Desert—were deposited over the Kaibab Formation. Cenozoic sediments followed, including gravel, sand, and volcanic ash deposited in a large inland lake.

With uplift of the Colorado Plateau, erosion stripped away both Mesozoic and Cenozoic deposits. With the growth of the San Francisco Mountain volcano, streams that wound across the plateau surface were strengthened and able to entrench their former meandering courses, some of them following the routes of collapsed caverns in the limestone plateau.

Prehistoric people of the Sinagua culture moved into the canyon 750 to 800 years ago, building one-story rooms with blocks of Kaibab limestone mortared with clay. The move to Walnut Canyon seems to have been part of a population shift that followed the eruption of Sunset Crater, dated by tree-ring studies as having begun in the winter of 1064/1065 A.D. Volcanic cinders from the eruption provided good mulch for Sinagua farms on the plateau surface, particularly around open "parks" and along shallow drainages, making them attractive to the agricultural people. Dwellings in Walnut Canyon were abandoned, though, before 1300 A.D., possibly because of a prolonged drought from 1276 to 1299, a drought that also left its traces in tree rings in this area.

Wupatki National Monument

Established: 1924
Size: 143 square kilometers (55 square miles)
Elevation: 1494 meters (4900 feet) at visitor center
Address: Tuba Star Route, Flagstaff, Arizona 86004

STAR FEATURES

• Prehistoric villages scattered along the edge of the Coconino Plateau, where Paleozoic sedimentary strata tilt and dive beneath Mesozoic rocks of the Painted Desert.

• Views of the broad valley of the Little Colorado River, part of Arizona's Painted Desert, and of San Francisco Mountain, monarch of the San Francisco volcanic field.

• Evidence of volcanism and of the effect it had on prehistoric agricultural people.

• Unusual earthcracks that may have been produced by earthquakes associated with the eruption of nearby volcanoes.

• Visitor center, museum, trails (with guide leaflets) to the ruins, conducted tours.

SETTING THE STAGE

West of the Little Colorado River and northeast of the San Francisco volcanic field, the Black Point Monocline defines the edge of the Coconino Plateau. Here, Paleozoic sedimentary rocks bend downward and plunge beneath Mesozoic rocks of the Painted Desert. Close to the monocline are ruins of several "apartment-house" villages dating back some 900 years, built by people of the Sinagua and Anasazi cultures in response to a geologic event: the eruption of Sunset Crater.

Like other ruins in this area, Wukoki Pueblo was constructed with blocks of red Moenkopi siltstone broken along natural joints. Sandstone edging its small mesa shows honeycomb weathering.

Lomaki Pueblo rests on the brink of a large earthcrack. When the pueblo was occupied, dwellers built granaries and storage rooms within the earthcrack. But a "dam" may have been built by modern Navajo to corral sheep.

Ruins of these villages give an exciting glimpse into geology's influence on prehistoric man.

In the Wupatki area, the Kaibab Formation, hard, buff-colored Permian rock that surfaces much of the Coconino Plateau, dips fairly gently eastward. This formation is made up of silty and sandy limestone and limy siltstone and sandstone. It contains pockets of marine fossils—mostly tiny clams and snails but a few large, thick-shelled forms as well. Its buff color makes it easy to recognize.

The Moenkopi Formation above it adds a brick red color to the national monument area. This rock unit consists of red mudstone, siltstone, and sandstone, some thin bedded and slabby, some thick bedded and massive. Many layers are cross-bedded; others are marked with the tracery of mudcracks, ripple marks, and salt-crystal impressions.

A number of lava flows from the San Francisco volcanic field extend into and across the national monument, two of them reaching the Little Colorado River. These rocks add the colors dark purple and black. Several purplish gray cinder cones are present in the monument as well, and the ground in many places is sprinkled with dark gray cinders from Sunset Crater. (For more on volcanism in this area, see Sunset Crater National Monument.)

Stresses that developed as the Coconino Plateau rose and the Black Point Monocline took shape

Citadel Pueblo builders blended irregular lumps of basalt with rectangular slabs of red sandstone.

fractured both the Kaibab and the Moenkopi Formations. Joints are at almost right angles and mark the rock into rectangular slabs, a convenience for Sinagua and Anasazi stonemasons, who worked, we must remember, without metal tools. Most of the ruins here are constructed from bricklike slabs of Moenkopi siltstone and small boulders and cobbles of basalt.

Of special interest in this national monument are a number of large earthcracks thought to have opened up during earthquakes associated with development of the San Francisco volcanic field. Some of the cracks act like blowholes, with steady breezes blowing from them at times when air pressure outside is falling, or with outside air blowing into them when air pressure is rising. Measurements of the amount of air "breathed" in and out show that these cracks communicate with extensive networks of underground crevices. Other cracks are partly filled with fine silt, and after a rain hold pools of water. Some of the Sinagua and Anasazi villages were built near such earthcracks.

GEOLOGIC HISTORY

Paleozoic Era. Paleozoic history here is the same as that of other Plateau areas: mostly marine sedimentation, with dunes forming in mid-Permian time. The only Paleozoic rock units appearing at the surface in the Wupatki area are the Kaibab Limestone and the Coconino Sandstone, both formed in Permian time in and near a shallow sea that came into this area from the west. Here the limestone is quite sandy, reflecting its nearness to the Permian shoreline. Its tiny fossil mollusks may have clung to marine grasses, while larger mollusks developed thick shells as protection against wave surge.

Mesozoic Era. Dark red siltstones and sandstones of the Moenkopi Formation are products of floodplains and coastal mudflats on a broad western continental slope. In several areas near the national monument, this formation contains fossils of large-bodied, flat-headed amphibians that swam or crawled about on Triassic floodplains and marshes. In places, tracks of a proto-dinosaur, *Chirotherium*, have been found. These tracks, shaped like human handprints with the "thumb" on the outside instead of the inside, were impressed into muddy surfaces and later covered with layers of sand.

After the Moenkopi Formation was deposited, a thin but remarkably widespread layer of gravel, with rounded pebbles of hard Precambrian rock, sheeted across this area. Today the gravel makes up the Shinarump Conglomerate, a resistant unit that caps a number of mesas within the national monument. Younger Triassic rocks, some of them documenting volcanic explosions somewhere to windward, appear in the Painted Desert to the east.

Cenozoic Era. Detailed studies of lava flows and cinder cones of the San Francisco volcanic field, and of terraces along the Little Colorado River, recount the Cenozoic history of this area.

Ten million years ago, long after the establishment of the Colorado Plateau and its subdivisions, the Little Colorado River roughed out the wide valley we see here today. Volcanic eruptions west of this valley began about 6 million years ago as lava flows spread over the area. At times, earthquakes shook this region as faults displaced the plateau-capping basalts. In some places, earthcracks opened up.

Periodically, lava flows from the volcanic field to the west intercepted the river, deflecting it from

A sink near Citadel Ruin formed along a fault that offsets limestone layers of the Kaibab Formation. A lava flow caps the background mesa.

its course. Erosion by the Little Colorado River between 4 and 2 million years ago, and eruption of the Black Point lava flow about 2.4 million years ago, created two river terraces 200 and 130 meters (650 and 425 feet) above the present river. Meantime, the great stratovolcano of San Francisco Mountain was slowly growing by additions of layer after layer of thick, silicic lava and volcanic ash. As the magma chamber below the volcano was depleted, central portions of the mountain collapsed into the void.

Two more erosion cycles 500,000 years ago left terraces 75 and 50 meters (246 and 164 feet) above the Little Colorado's present bed.

During Pleistocene time, glaciers developed on San Francisco Mountain. A glacier in the Inner Basin, the collapsed center of the mountain, contributed outwash gravels that extend all the way to the Wupatki area.

Then, 150,000 to 80,000 years ago, renewed erosion reshaped the valley of the Little Colorado River and created terraces 25, 15, and 9 meters (80, 50, and 30 feet) above the present bed of the river. Since then, the Little Colorado River has cut down about 10 meters (30 feet) into the gravel deposits of the lowest terrace.

Agricultural Sinagua and Anasazi farmers came into this area about a thousand years ago. At first they built small pit houses near their farms. When Sunset Crater erupted in 1064/1065 A.D., they probably fled, but they later returned to find that the layer of cinders scattered by the eruption served as a natural mulch, greatly improving the productivity of their fields. Tree-ring studies indicate that the climate may have been cooler and moister for several years following the eruption, an additional factor in promoting successful farming.

The news apparently spread and brought in peoples of other cultures as well. Pueblos were built at Wupatki, Wukoki, and other sites to take advantage of the abundant crops. Obtaining water from a few springs and seeps, catching and saving rainwater, the farming people remained here until about 1250 A.D. Then they abandoned their villages for reasons as yet uncertain—possibly because of depletion of soil and sources for firewood or because of subtle climate changes. They may have fused with their pueblo-building relatives, the Hopis, who live northeast of here, or they may have migrated to the Verde Valley southwest of here, where the Verde River provided a more dependable water supply.

OTHER READING

Schroeder, Albert H., 1977. *Of Men and Volcanoes—the Sinagua of Northern Arizona.* Southwest Parks and Monuments Association, Globe, Arizona.

Zion National Park

Established: 1909 as Mukuntuweap National Monument, 1919 as Zion National Park
Size: 593 square kilometers (229 square miles)
Elevation: 1117 to 2664 meters (3666 to 8740 feet)
Address: Springdale, Utah 84767

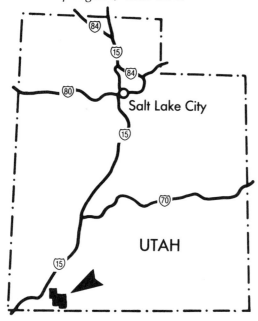

STAR FEATURES

• A dramatic canyon with towering walls of colorful Triassic-Jurassic sandstone, relic of an ancient desert.

• The North Fork of the Virgin River, carver of the canyon—placid much of the year but often rising to furious pitch during the summer rainy season.

• Smaller canyons, deep clefts between narrow rock ramparts, many controlled by parallel joints and faults.

• Stone arches, freestanding or still attached to the cliffs behind.

• Abundant evidence of rockfalls and slides, and a line of springs that contribute to them.

• Evidence of Quaternary volcanism.

• Visitor centers, museum, hiking trails (one with a guide leaflet), guided walks and evening programs in summer. A geologic map of Zion is available at the visitor centers.

See frontispiece and color section for additional photographs.

SETTING THE STAGE

Deeply carved in the southern end of the Markagunt Plateau, Zion Canyon boasts some of the highest sheer cliffs in America, a few of them 600 meters (2000 feet) from top to base. The soaring monuments that confine the canyon, painted in delicate hues of pink and white, are shaped in a major Plateau scenery-maker, the Navajo Sandstone. The sweeping cross-bedding of this formation lends style to the canyon walls.

In Zion the Navajo Sandstone is 600 meters (2000 feet) thick; elsewhere it reaches 1000 meters (3000 feet). And it extends far beyond Zion, from central Wyoming (where it is called the Nugget Sandstone) to southern Nevada. Over all of this area it is remarkably similar in texture and composition, being composed of fine quartz sand with variable amounts of iron oxide, which give it its pink color, and calcite. Wherever its grains are cemented together with silica, the sandstone is strong enough to stand in tall vertical cliffs.

What factors led to the building of this great body of sandstone? Its large-scale, concave-upward cross-bedding and uniformly fine, rounded and frosted sand grains show that it accumulated as dunes on a windswept desert. Its great areal extent tells us that this desert once covered a region nearly as large as the Sahara. Horizontal siltstone layers between the slanting sets of cross-beds show us that silt and clay particles accumulated in flat interdune areas like those known from the Sahara and other modern deserts.

Gradual downcutting by the North Fork of the Virgin River shaped the Zion Canyon of today. Admittedly, the North Fork usually seems too clear, too placid to have accomplished all this erosion. But with summer thunderstorms or spring melting of snow on the Markagunt Plateau around its headwaters, the river picks up volume and speed, and pounds its bed with sand, pebbles, and even large boulders, continuing to cut down at an average rate of about 2 centimeters (1 inch) per century, roughly 200 meters (600 feet) per million years. Fed by the more abundant snowfall and rainfall of Pleistocene time, downward cutting may have far exceeded this figure.

Canyon widening is the job of slides and rockfalls. The part of Zion Canyon that is accessible by road is edged with tumbled blocks and rockslide material. The Kayenta Formation, below the Navajo Sandstone, is responsible for most slides; its relatively soft siltstones and mudstones fairly invite erosion. As it wears away, the Navajo Sandstone is undermined, and despite its inherent strength eventually breaks away, sometimes along vertical joints, sometimes in large curving slabs.

Sweeping cross-bedding in the Navajo Sandstone is a relic of a Triassic-Jurassic desert as extensive as the modern Sahara. Tad Nichols photo.

Contorted bedding such as that shown here originated in slumps or sand avalanches on ancient dunes. Tad Nichols photo.

In the Kolob section of Zion National Park, trails wind into a wilderness of narrow canyons and steep coral-colored cliffs. Ray Strauss photo.

Proof of the Kayenta Formation's role in canyon widening can be found in the Narrows, beyond the end of the paved trail. There, where the river has not yet cut down to the Kayenta Formation, Navajo Sandstone cliffs come right to the water's edge. But just below the Narrows, near the end of the road, the Kayenta Formation occurs low down on the canyon walls. A recent rockslide partly dams the stream, its giant boulders giving evidence that canyon widening is taking place.

Undermining and collapse are abetted by a line of springs along the contact between the Navajo Sandstone and the Kayenta Formation. Rainwater and snowmelt that filter down through the permeable sandstone eventually meet the impermeable mudstones of the Kayenta Formation, and are turned aside along the contact between the two formations. Where the contact is exposed, the water emerges as a line of springs low down on the canyon walls.

The spring line at the base of the Navajo Sandstone, as well as smaller ones along interdune deposits, are also responsible for some of Zion's great arches. (Unlike those of Arches National Park, most of Zion's arches are not freestanding, but are really large alcoves in the canyon walls.) Arches develop where there are relatively few vertical joints, where undermined Navajo Sandstone breaks away along curving fractures. Overhangs like that at Weeping Rock form the same way but on a smaller scale. Near springs, water in narrow cracks expands as it freezes on winter nights, helping to break the rock apart.

Where rock breaks along vertical joints, it also influences patterns of erosion. Some of the tributaries of the North Fork, as well as many tributaries of tributaries, follow the north-northwest trend of faults and joints that slash the strata here, giving an overall fabric to the land. The trend of these joints and faults is parallel to the trend of the Hurricane Fault, the major fault that separates the Colorado Plateau from the Basin and Range region to the west. Along the Hurricane Fault, offset is as great as 1200 meters (4000 feet). A further offset of up to 150 meters (500 feet) is provided by the East Cougar Mountain Fault, which cuts through eastern and northeastern parts of the park.

Details of weathering of the Navajo Sandstone are everywhere visible in the park. Wind etches the cross-bedding, desert varnish adds dark peacock tones to rock surfaces, black lichens stripe canyon walls, seeping water leaves white streaks of calcium carbonate, and growing plants pry rocks apart.

GEOLOGIC HISTORY

Mesozoic Era. Since Paleozoic rocks are exposed only in the northwest tip of the national park, where a tiny outcrop of Permian marine limestone comes to the surface along the Hurricane Fault, we'll begin our history with events of Mesozoic time.

As the Triassic Period began, this part of North America had just risen above the sea and formed a broad coastal plain sloping westward toward the ocean. Sluggish rivers wound across the plain, gradually depositing silt, mud, and clay—derived from the Ancestral Rocky Mountains—to form the Moenkopi Formation. As sea level fluctuated, beds of limestone accumulated and isolated lagoons of seawater evaporated, leaving layers of salt and gypsum. Lying close to the equator, the region was tropical: hotter and more humid than it is today.

In mid-Triassic time, the nature and probably the source of the sediments changed. For a time, coarse sands and gravel were deposited in a thin but widespread sheet, to form the Shinarump Conglomerate. Pebbles in this unit seem to have come from highlands in central Arizona. Still later these coarse sediments were covered with fine volcanic ash and with mudflows that may have been initiated by volcanic eruptions. These layers, purple and green and gray, contain fossil amphibians and reptiles as well as petrified wood; they make up the Chinle Formation. Coarser river channel and floodplain deposits of the Moenave and Kayenta Formations were then deposited above the Chinle Formation, indicating cessation of volcanism in late Triassic time.

During the Mesozoic Era, North America slowly drifted westward and northward. By late Triassic time it had entered the dry subtropical latitudes, then as now the site of major deserts. On a wide, windswept Triassic-Jurassic plain, sand dunes piled up, layer upon layer, for millions of years. This sea of sand, stretching from northern Arizona to southern Wyoming, is now the Navajo Sandstone.

A few thin layers of red siltstone, the Temple Cap Formation, and above it thinner cross-bedded sandstones of the Carmel Formation, occur above the Navajo Sandstone. These rocks show that the western sea flooded this area briefly near the end of the Jurassic Period. Much later, in Cretaceous time, the sea again advanced, this time coming from the east. The Dakota Sandstone, formed on its beaches, is the youngest sedimentary unit in Zion National Park; it appears only on the lava-protected summit of Horse Ranch

Mountain, in the northern part of the park. Marine sediments—fine gray shales, sandstones, and limestones—fill out the Cretaceous sequence elsewhere on the Plateau, but do not appear in Zion.

Near the end of the Mesozoic Era, the Rocky Mountains rose east and north of here, stretching in a great arc from Canada to northern New Mexico. The rise of the mountains brought an end to the pulsing incursions of the sea.

Cenozoic Era. As the Rockies rose, deep intermountain basins in Utah and Wyoming began to receive rock debris washed from them. Lakebeds deposited in one of these basins now show up in Bryce Canyon National Park and Cedar Breaks National Monument, but in the Zion area all Tertiary deposits have been eroded away.

Zion does, however, contain evidence of Quaternary lakes, their waters dammed by lava flows or by rockslides within the canyon. Lava flows and cinder cones along the west edge of the park show that volcanic magma rose along the faults that edge the Colorado Plateau, faults deep enough to tap the basalt magma of the Earth's mantle. The resulting volcanic rocks have been dated as 260,000, 500,000, and 1.4 million years old.

The primary geologic process of Quaternary time, however, was (and is) erosion. The amount of canyon cutting suggests that erosion was faster in Pleistocene time, when precipitation was greater than at present. Rockfall debris that now forms benches above the lower part of Zion Canyon and Coalpit Wash was formed in Pleistocene time.

About 4000 years ago a rockslide occurred at the lower end of Zion Canyon, 2 kilometers (1 mile) above the bridge and the junction of the highway routes into Zion. Damming the North Fork, the slide created a long, narrow lake that lasted for about 1500 years. Remains of the rockslide jut into the canyon as a jumble of immense boulders and rocks with no apparent stratification, and sediments deposited in the lake now form the level area around Zion Lodge. Red Arch, near Refrigerator Canyon, fell in historic time, obliterating a cornfield and a spring belonging to a local farmer.

BEHIND THE SCENES

Angels Landing. The trail to Angels Landing offers unusual views of both upper and lower Zion Canyon. Here the North Fork is deflected from a southward course into a tight loop around Angels Landing and the Organ. There are also good views of the bold north face of the Great White Throne, capped (as are parts of Zion's East and West Rims) with thin layers of the Temple Cap Formation. Details of the Navajo Sandstone are visible along the trail. Tributary streams such as that in Refrigerator Canyon follow the north-northwest trend of parallel joints, a feature easy to observe from the summit.

Court of the Patriarchs. Standing at the west side of Zion Canyon, the Patriarchs display the large-scale cross-bedding and pink and white colors of the Navajo Sandstone. North-northeast-trending joints divide the Patriarchs and control the erosion of nearby rock clefts.

The perfectly flat canyon floor below the Patriarchs is underlain by lake deposits formed when a rockslide dammed the river.

Checkerboard Mesa. Here, erosion by wind and rain has creased the surface of the Navajo Sandstone. Rivulets have incised small gullies along vertical joints, while wind and blowing sand have etched the sweeping cross-bedding of this rock. Checkerboarding seems to be limited to north-facing slopes. Like mesas north of the highway, Checkerboard Mesa is capped with striped layers of the Temple Cap and Carmel Formations. Several arches in the Navajo Sandstone can be seen north of the highway.

Coalpit Wash. Geologically interesting, this small valley contains Pleistocene lake deposits—clay, silt, and limestone—and broken volcanic rock that form a bench north of the lava flow that dammed the valley about half a million years ago. The lava is associated with two cinder cones together called Crater Hill. The lake deposits contain a fossil camel track, as well as fossil pollen that shows the climate to have been moister in Pleistocene time than it is at present.

East Rim Trail and Observation Point (5.8 kilometers, or 3.6 miles). Once the only route from Zion Canyon to communities farther east, the trail up Echo Canyon passes through a narrow slot formed by merging potholes in the stream. Such potholes take shape as boulders and pebbles—natural grindstones—are whirled in slight depressions in the stream bed. As walls between successive potholes were ground away, the stream rushed through the deep, twisting chasm.

Above this narrow passage the canyon opens out, and the trail is surrounded with cliffs and steep slopes of Navajo Sandstone. Cross-bedding in the Navajo Sandstone is particularly prominent here, with long, slanting cross-beds separated by

Rocky tree-covered talus slopes below Zion's perpendicular walls bear testimony to many rockfalls. The West Rim Trail, at the bottom of the photograph, climbs the talus slope. Red Arch Mountain, site of a fairly recent rockfall, is at the upper left. Ray Strauss photo.

The trail to Angels Landing ascends the narrow ridge at the right. Joints control the shape of cliff faces, which are painted with desert varnish and black lichens. Ray Strauss photo.

Cross-bedding combines with vertical fractures to give an unusual appearance to Checkerboard Mesa. Tad Nichols photo.

flat interdune deposits. Erosion along joints, desert varnish, lichens, red stain washing down from the Temple Cap Formation, and many other geologic features can be seen here.

The pink color of the lower part of the Navajo Sandstone is caused by tiny particles of hematite (iron oxide) within the formation. The color may mark a stable groundwater level long before Zion Canyon was carved. Elsewhere the Navajo Sandstone varies from white to yellow to entirely pink.

The trail eventually climbs into the Temple Cap and Carmel Formations above the Navajo Sandstone. Following the ledge at the top of the Navajo Sandstone, the trail reaches Observation Point, with a magnificent view of the Great White Throne, Angels Landing, and almost all of Zion Canyon.

Gateway to the Narrows. In upper parts of Zion, the canyon of the North Fork becomes so narrow that the river occupies its entire width. The Kayenta Formation is well below river level, and cliffs of Navajo Sandstone rise sheer and tall from the water's edge. Summer floods rushing through the Narrows fill the gorge from wall to wall.

At the Gateway to the Narrows, where the Kayenta-Navajo contact is just above river level, springs along the contact turn woodlands into swamps. Two recent rockslides here partly dammed the river, tumbling giant boulders into its course. Much of the rockslide is a mixture of sand and mud—broken up rock—but there are also plenty of larger rock fragments. The scar from which the slides came can be seen well up on the canyon wall.

The view from the West Rim reveals the flat upper surface of the Markagunt Plateau, part of it the top of the Great White Throne (left half of photograph). Ray Strauss photo.

Elsewhere the cliffs above the river are marked with streaks of black lichens, iridescent coats of desert varnish, and patches of white calcium carbonate leached from the rock itself. High-up red streaking is fine mud from the Temple Cap Formation, which overlies the Navajo Sandstone. Note the joint patterns in the Navajo Sandstone, and the tendency of this unit to break away in arches.

Boulders in the stream are representative of the country around this river's headwaters. Some are of dark gray basalt, the rock that caps most of the Markagunt Plateau. Across the river are several freestanding pinnacles of Navajo Sandstone, perched precariously on pedestals of Kayenta Formation siltstone that eventually will be worn away by the river.

Hanging Gardens. Along the line of the Navajo-Kayenta contact, and on a smaller scale along horizontal interdune deposits, spring water seeps from the Navajo Sandstone. The water's source is in rainfall and snowmelt on the surface of the Markagunt Plateau. Trickling downward through the porous sandstone, water is forced to flow sideways by relatively impervious mudstones of the Kayenta Formation or by similarly impervious interdune deposits. Wherever it then intercepts the canyon, it comes forth in springs that nourish hanging gardens of water-loving plants.

The steadily seeping water weakens surrounding rock by repeatedly freezing in winter and by dissolving the cement that holds sand and silt grains together. As a result, springs are commonly shadowed by overhanging ledges.

Hidden Canyon. Hidden Canyon is one of the many deep, narrow, almost inaccessible canyons that cut the southern Markagunt Plateau, deep clefts eroded along north-northwest trending joints. The canyon is about 1.5 kilometers (1 mile) long and at the most only 20 meters (65 feet) wide. Vertical or overhanging walls of Navajo Sandstone are marked with lichens and stained with desert varnish. Here and there are little grottos, the work of water and wind. A small natural bridge is present within the canyon as well.

Kolob Canyons. The unfrequented northwestern part of Zion National Park is no less attractive than Zion Canyon. The road follows Taylor Creek into this part of the park, traveling on Triassic rocks of the Moenkopi, Chinle, and Moenave Formations. Turning southeast across a landslide, it leaves Taylor Creek and crosses a thrust fault where Jurassic rocks have pushed westward over the Triassic units.

Except near the South Fork of Taylor Creek, the road then remains on the Moenave Formation. Views eastward show the long buttresses of the Kolob Plateau, walled with Navajo Sandstone and capped with reddish ledges of the Temple Cap and Carmel Formations. The buttresses are separated by the Finger Canyons of the Kolob Plateau. Several trails wind into this wilderness, some leading to rock arches, among them Kolob Arch, its 94-meter (310-foot) span one of the longest known natural spans in the world. In dry weather the sandy floors of the Kolob canyons make enjoyable trails, too.

Kolob Reservoir Road. Entering the national park on some of the lava flows that mark its western margin, this road provides easy access to some of Zion's volcanic features. The magma of the lava flows and cinder cones rose along West and East Cougar Mountain Faults. Its basalt composition tells us that these faults extend down through the Earth's crust to the mantle. The molten rock flowed down a former valley of North Creek and its tributary Grapevine Wash, displacing the streams. As surrounding rocks wore away, the resistant lava remained as a ridge.

Traveling north between the two Cougar Mountain Faults, where flat-lying rocks tilt eastward in response to the fault movement, the road passes rows of small domes, the larger Tabernacle Dome, and hoodoos and cliffs of white Navajo Sandstone.

Spendlove and Firepit Knolls are cinder cones, with craters at their summits. From the rim of Firepit Knoll—a short hike—vistas extend southward to the great cliffs of Zion Canyon, northward to the ragged edges of the Kolob Plateau, and eastward over more lava flows to parts of the Markagunt Plateau that surround the headwaters of the North Fork. To the west are the Basin and Range deserts of southwestern Utah. The East Cougar Mountain Fault runs between the Spendlove and Firepit cinder cones.

Petrified Forest. In the Chinle Formation in the southwest corner of the park, accessible by trail, Zion's Petrified Forest contains fossil logs similar to those in Arizona's Petrified Forest, which is in the same formation. Many of the logs occur in the Shinarump Conglomerate at the base of this formation. In places, cavities within the logs contain the bright yellow uranium ore, carnotite.

Weeping Rock. Rain and snowmelt on the Markagunt Plateau seep down through younger rock units and into the porous Navajo Sandstone. Prevented from further downward movement by

Zion's trails lead through narrow canyons and rugged uplands of bare rock, where the Navajo Sandstone is the major scenery-maker. Ray Strauss photo.

The highway to Mount Carmel climbs a landslide before disappearing into the tunnel to ascend through the Navajo Sandstone. Ray Strauss photo.

fine-grained Kayenta Formation mudstone and siltstone, they eventually emerge as springs along the contact between the two formations. As elsewhere along this spring line, water-loving plants ornament the site. Because the water weakens the rock, making it more susceptible to erosion, some of the rock has fallen away here, leaving an arching overhang.

Zion–Mount Carmel Highway and Tunnel. Completed in 1930, the highway travels up Pine Creek Canyon and tunnels through the otherwise insurmountable cliffs of Navajo Sandstone that confine Zion Canyon.

In this part of Zion the Moenave and Kayenta Formations form steep, ledgy slopes below the Navajo Sandstone cliffs. The highway zigzags up this slope and over a large but seemingly stable landslide (with views of the Great Arch on the opposite canyon wall), to the tunnel's western portal, right at the Kayenta-Navajo contact. A short trail near the east portal leads to a viewpoint and panoramic vistas of Pine Creek Canyon and lower Zion Canyon.

East of the tunnel the highway winds through bare rock outcrops of Navajo Sandstone. Many side canyons parallel the north-northwest joint pattern prevalent in this area, well revealed on Checkerboard Mesa.

OTHER READING

Breed, W.J., 1983. *Geologic Cross-section of the Cedar Breaks–Zion–Grand Canyon Region.* Zion Natural History Association, Springdale, Utah.

Eardley, A.J., and Schaack, J.W., 1985. *Zion: the Story behind the Scenery.* KC Publications, Inc., Las Vegas, Nevada.

Grater, R.K., and others, 1978. *Zion.* Unicorn Associates, Denver, Colorado.

Hagood, Allen, 1985. *This is Zion.* Zion Natural History Association, Springdale, Utah.

Hamilton, W.L., 1984. *Geologic Map of Zion National Park.* U.S. Geological Survey.

Hamilton, W.L., 1984. *The Sculpturing of Zion.* Zion Natural History Association, Springdale, Utah.

Glossary

aa lava—extremely rough, fragmented lava.

agate—a banded or otherwise decorative variety of quartz.

alluvial—deposited by rivers and streams.

amethyst—a lavender variety of quartz.

ammonite—an extinct, shelled relative of modern squids, octopuses, and chambered nautilus.

andesite—a medium gray, silicic volcanic rock.

anticline—a fold in layered rocks that is convex upward.

arch—a stone arch formed by erosion of rock, not bridging a watercourse.

archaeology—the study of prehistoric man's life and culture.

arroyo—a gully or small canyon, dry most of the time.

ash—fine particles of pulverized magma blown from a volcanic vent.

badlands—rough, gullied topography in arid and semiarid regions eroded by infrequent but heavy rains.

basalt—dark gray to black volcanic rock poor in silica and rich in iron and magnesium minerals.

basin—a downwarped or downdropped area filled with sediment eroded from surrounding higher areas.

bedding—layering of sedimentary rocks.

bedrock—solid rock exposed at or near the surface.

bench—a wide, gentle slope partway up a slope or canyon wall.

bentonite—a soft, porous, light-colored clay mineral formed by decomposition of volcanic ash.

biotite—black mica

bomb, volcanic—a fragment of molten or semi-molten rock thrown from a volcano.

boulder—a large, rounded rock fragment with a diameter greater than 25 centimeters (10 inches), usually transported by running water.

brachiopod—a marine shellfish with two bilaterally symmetrical shells, common as fossils in Paleozoic rocks.

breccia (rhymes with "betcha")—rock consisting of coarse, broken rock fragments imbedded in finer material such as volcanic ash.

bridge, natural—a freestanding rock span created by a stream eroding and finally penetrating a rock spur.

bryozoan—a group of branching, coral-like invertebrates.

butte—a steep-walled hill capped with resistant rock.

calcite (calcium carbonate)—$CaCO_3$, major mineral component of limestone and travertine, also occurring as cementing material in sandstone and siltstone.

caliche (ca-LEE-chee)—whitish calcium carbonate-cemented gravel found on or near the surface in arid and semiarid climates, deposited as calcium carbonate-bearing water is drawn to the surface and evaporated.

caprock—a resistant rock forming the top of a butte, mesa, or plateau.

cavern—a large and usually complex cave.

cephalopod—a group of marine mollusks with tentacles, including octopi, squids, and nautilus as well as extinct ammonites.

chert—a hard, dense form of silica that usually occurs as nodules in limestone.

cinder cone—a small, conical volcano built primarily of loose fragments of bubbly volcanic material thrown from a volcanic vent.

cinders—bubbly, popcornlike volcanic material.

clay—very fine rock material, with particle sizes of less than 4 microns (0.00016 inch) in diameter.

cobble—a rounded rock fragment having a diameter of 6.4 to 25 centimeters (2.5 to 10 inches).

composite volcano—a volcanic mountain built of layered lava, volcanic breccia, and volcanic ash; a stratovolcano.

conchoidal fracture—a smoothly curved fracture such as that occurring in broken glass.

concretion—a rounded mass of mineral matter found in sedimentary rock.

conglomerate—rock composed of rounded, water-worn fragments of older rock.

contact—the boundary between two rock units.

continental—sedimentary rocks deposited on land or in lakes, by streams, wind, or ice.

coral—a group of sea-dwelling animals that may deposit calcium carbonate in large reeflike masses.

crater—the funnel-shaped hollow at the summit of a volcano, including the vent from which lava and volcanic ash are ejected.

crinoid—a marine animal related to starfish, having a lilylike body on a long, segmented stem.

cross-bedding—slanting laminae within a sedimentary rock layer.

crust—the outermost shell of the Earth, above the mantle and core.

cuesta—a ridge with a long, gentle slope formed by a tilted, resistant rock layer, and a short, steep slope on the cut edges of that and lower layers.

cyclic—occurring in cycles.

dendritic drainage—a treelike pattern of branching streams and rivulets.

desert pavement—a natural concentration of closely packed pebbles, the result of winnowing by wind action and sheetwash.

desert varnish—a thin, glossy coating of dark brown or blue-black manganese and iron minerals on rocks in desert regions.

differential erosion—irregular erosion caused by differences in rock hardness or resistance.

dike—a sheetlike intrusion that cuts vertically or nearly vertically across other rock structures. In igneous rocks, dikes are often called **veins**.

dip—the direction and degree of tilt of sedimentary layers, measured downward from horizontal.

displacement—offset along a fault or monocline.

dissected—cut into by stream erosion.

dolomite—a sedimentary rock consisting of calcium and magnesium carbonates.

dome—a more or less circular anticline in which rocks dip away in all directions. (See also **lava dome**.)

drag—bending of rock layers near faults, caused by friction along the fault surface.

earthcrack. An open crack caused by an earthquake.

eolian—caused by wind.

entrenched—occupying a trench cut by stream or river erosion.

epidote—a yellowish or greenish mineral found in metamorphic rocks.

epoch—a unit of geologic time, subdividing a period.

era—the largest unit of geologic time.

erosion—the process by which rock is loosened, dissolved, and worn away.

evaporite—minerals left behind by evaporation of sea or lake water; includes salt, gypsum, potash, and anhydrite.

exfoliation—a process in which concentric crusts of rock break away from a rock surface. Also called **spalling**.

extrusive igneous rock—rock formed of magma that flows out on the surface and solidifies there. Also called **volcanic rock**.

fault—a rock fracture along which displacement has occurred.

fault block—a segment of the Earth's crust limited on two or more sides by faults.

feldspar—a group of common, light-colored, rock-forming minerals containing aluminum oxides and silica.

floodplain—nearly horizontal land adjacent to a river channel, with sand and gravel layers deposited by the river during floods.

fold—a curve or bend in rock strata.

formation—a mappable unit of stratified rock.

fossil—remains or traces of a plant or animal preserved in rock; also, long-preserved inorganic structures such as fossil ripple marks.

frost heaving—lifting of rocks or soil by crystal expansion as water in them freezes.

frost wedging—prying apart of rock by crystal expansion as water freezes repeatedly in cracks and crevices.

fumarole—a vent through which volcanic gases or vapors are emitted.

glaciation—presence of or erosion by glaciers.

glacier—a large mass of ice driven by its own weight to move slowly downslope or outward from a center.

glauconite—a green mineral containing iron, thought to form in agitated sea water.

gneiss (pronounced "nice")—banded metamorphic rock formed from granite (which it commonly resembles), sandstone, and other continental rocks.

gooseneck—a curve of an entrenched stream or river course.

graben—a linear valley dropped down between two parallel or nearly parallel normal faults.

gradient—the angle of slope of a stream or river course.

granite—a granular intrusive rock composed primarily of crystals of quartz and feldspar, peppered with dark biotite and/or hornblende crystals; any light-colored, granular intrusive rock.

gravel—a mixture of pebbles, boulders, and sand not yet consolidated into rock.

groundwater—subsurface water, as distinct from rivers, streams, seas, and lakes.

group—a major rock unit consisting of two or more formations having certain characteristics in common.

gypsum—a common evaporite mineral, calcium sulfate; $CaSO_4.2H_2O$.

hardpan—a relatively hard or impervious soil layer in which soil particles are cemented together by minerals added by soil water.

headward erosion—stream erosion in an upstream direction, particularly at the head of a gully or canyon.

hematite—a common dark red, iron oxide mineral; Fe_2O_3.

hogback—a steep, sharp-crested ridge with approximately equal slopes, one formed of a hard, steeply tilted caprock, the other of the edges of the caprock and layers below.

honeycomb weathering—weathering of sandstone into small, deep pits, usually by wind.

hoodoo—a grotesque pinnacle or pillar of rock.

hornblende—a black or dark green mineral whose rodlike crystals are common in igneous rocks.

humus—decomposed plant material in soil.

Ice Ages—the Pleistocene Epoch. (See geologic calendar, p. 18.)

igneous rock—any rock formed from molten magma.

impermeable or **impervious**—not allowing penetration of water or other fluids.

interdune deposit—a silty or clayey deposit formed in a low, flat area between sand dunes.

intrusion—an igneous rock mass formed from molten magma that does not reach the surface.

intrusive rock—igneous rock created as molten magma intrudes pre-existing rocks and cools without reaching the surface.

iridium—a chemical element rare in surface rocks, more common in meteorites and the Earth's mantle.

jasper—a dense, opaque, often colorful variety of chert, a variety of quartz.

joint—a rock fracture along which no significant movement has taken place.

karst—a distinctive type of rough, irregular landscape formed by solution of limestone.

laccolith—a dome due to forceful injection of magma between sedimentary layers, doming upper layers.

lagoon—quiet near-shore water sheltered by a reef or offshore bar.

laminae—thin layering of rock visible on eroded rock surfaces.

landform—recognizable features of the Earth's surface, such as hills, valleys, cliffs, mesas, and pinnacles.

Laramide Orogeny—mountain building of late Cretaceous and early Tertiary time, creating the Rocky Mountains.

lava—molten magma that has reached the Earth's surface, or the rock formed when such magma cools.

lava dome—a type of volcano characterized by very thick magma that piles into a rounded dome above its conduit.

lava flow—an outpouring of molten lava or a rock mass formed from it.

lichen—a plant community consisting of a fungus and an alga, appearing as a flat, often circular crust.

liesegang rings—concentric rings or bands formed as minerals precipitate in water-saturated rock.

lime—a term commonly, though incorrectly, used for calcium carbonate.

limestone—a sedimentary rock consisting largely of calcium carbonate.

limonite—a yellow-brown iron oxide mineral; $2Fe_2O_3.3H_2O$.

lithosphere—the outermost rocky layer of the Earth.

magma—molten rock. When extruded onto the Earth's surface, magma is usually called **lava**.

magma chamber—a reservoir of magma from which volcanic materials are derived, usually only a few kilometers below the surface.

magnetite—a black iron mineral having magnetic properties; $(Fe,Mg)Fe_2O_4$.

mantle—the thick, partly molten zone between the Earth's core and crust.

marine—formed in the sea.

meander—a loop on the sinuous course of a river.

mesa—a large flat-topped hill with a resistant caprock and steep slopes, larger than a butte but smaller than a plateau.

metabolic—having to do with life processes.

metamorphic rock—rock derived from preexisting rocks altered by heat, pressure, and other processes.

metamorphism—alteration of rock by heat, pressure, and chemical processes.

mica—a group of complex silicate minerals characterized by shiny, closely spaced, parallel layers that split apart easily.

mid-ocean ridge—a continuous, linear mountain range extending through the Earth's ocean basins, site of sea-floor spreading.

mineral—a naturally occurring inorganic substance with a characteristic chemical composition, color, texture, and crystal form.

mollusk—an animal group including clams, snails, octopi, squid, and several extinct forms.

monocline—a fold in otherwise horizontal sedimentary rocks that flexes in one direction only, with similar layers at different levels on either side.

mud cracks—shrinkage cracks in drying mud.

mudflow—a flowing mass of fine mud.

mudstone—rock formed from mud, with clay- and silt-sized particles.

natural bridge—a natural rock arch that spans the stream responsible for its formation.

normal fault—a fault in which the hanging (upper) wall moves downward relative to the footwall.

oceanic crust—part of the Earth's crust that occupies the sea floor and that originated along mid-ocean ridges.

outcrop—bedrock that appears at the surface.

overburden—rock material overlying a rock unit or mineral deposit.

pahoehoe lava—lava with a glassy, smooth or ropy surface.

paleomagnetic dating—dating of rocks by comparison with known patterns of reversal of the Earth's magnetic field.

pebble—a rock fragment, commonly rounded, 0.4 to 6.4 centimeters (0.2 to 2.5 inches) in diameter.

pegmatite—exceptionally coarse-grained igneous rock found as dikes or veins in large igneous intrusions.

peneplain—a level or almost level erosion-caused surface of fairly wide extent.

period—a subdivision of geologic time shorter than an era, longer than an epoch.

permeable—allowing penetration by water or other fluids.

petrified—turned to stone.

petroglyph—rock carving.

phytosaur—an extinct swimming reptile similar to the crocodile.

pinnacle—a fingerlike tower of rock.

plate—a block of the Earth's crust, separated from other blocks by mid-ocean ridges, trenches, and/or collision zones.

plateau—a flat-topped tableland more extensive than a mesa.

Plate Tectonics, Theory of—a theory that explains why sea floors spread and continents move apart as new crust is created at mid-ocean ridges.

pothole—a circular depression ground out by pebbles, cobbles, and sand swirled by running water, or formed by solution in small pools of rainwater.

pressure ridge—a transverse ridge in a lava flow, caused by pressure of continued flow.

pumice—light-colored, frothy, lightweight volcanic rock.

quartz—crystalline silica, SiO_2, a common rock-forming mineral.

quartzite—sandstone consisting chiefly of quartz grains welded firmly together with silica.

racetrack valley—a circular or oval valley around a mountain mass, shaped in the eroded edges of weak sedimentary rock layers that formerly arched over the mountain.

radiometric dating—dating of rocks by analysis of decay of radioactive minerals.

redbeds—predominantly reddish sedimentary rocks.

reverse fault—a fault in which the hanging (over-hanging) wall moves upward relative to the footwall.

rift valley—a large graben bordered by faults that reach through the Earth's crust to the mantle.

rimrock—resistant rock forming canyon or plateau rims.

ripple mark—a pattern of small ridges formed as water or wind transports and deposits sand or silt.

rockslide—a landslide involving a large proportion of rock.

salt anticline—an anticline pushed up by a rising mass of salt.

salt dome—a dome pushed up by a rising mass of salt.

sand—rock fragments 62-4000 microns (0.0025-0.016 inch) in diameter.

sandstone—sedimentary rock formed of sand-sized particles.

schist—crystalline metamorphic rock that tends to split along parallel planes.

scour marks—striations eroded in a rock face by streams or slides in an adjacent gully.

sea-floor spreading—movement of oceanic crust away from mid-ocean ridges, and creation of new oceanic crust at the ridges.

sedimentary rock—rock composed of particles of other rock transported and deposited by water, wind, or ice.

sediment—fragmented rock, as well as shells and other animal and plant material, deposited by wind, water, or ice.

selenite—a clear variety of gypsum commonly found in veins.

shale—fine-grained mudstone or claystone that splits easily along bedding planes.

silica—a hard, resistant mineral—SiO_2—that in its crystal form is quartz. It also occurs as opal, chalcedony, siliceous sinter, chert, and flint.

silicic—having a high proportion of silica.

sill—a flat igneous intrusion that has pushed between layers of stratified rock.

silt—rock fragments 4-62 microns (0.00016-0.0025 inch) in diameter.

siltstone—rock made of silt-sized particles.

sink or **sinkhole**—a more or less circular depression created when part of the roof of a cavern collapses.

slump—a landslide in which rock and earth slide as a single mass along a curved slip surface.

solution cavern—a cavern formed by solution of limestone.

solution valley—a valley formed by collapse of a solution cavern or by joining of a row of sinks.

spalling—peeling away of thin surface layers of rock.

spring line—a rock contact along which groundwater comes to the surface.

squeeze-up—a jagged mass of lava pushed up through a crack in an overlying lava crust.

strata—layers of sedimentary (and sometimes volcanic) rocks.

stratified—layered.

stratigraphic—pertaining to the study of strata (and therefore of sedimentary rocks).

stratovolcano—a volcanic mountain built of alternating layers of lava, breccia, and volcanic ash.

structure—pertaining to the form and arrangement of rocks, in particular to faults and folds.

subduction—the downward plunge of an oceanic plate below a continental plate.

syncline—a fold that is convex downward.

talus—a mass of large rock fragments lying at the base of a cliff or steep slope from which they have fallen.

terrace—a horizontal shelf along a river valley, formed when a river erodes down into its former floodplain.

travertine—a hard, dense limestone deposited by calcium-carbonate-laden water of streams, hot springs, and caves.

tree-ring dating—dating by comparing growth rings of trees or other wood.

trench—a sea-floor depression marking the zone where oceanic crust dives beneath continental crust.

trilobite—a crablike invertebrate that became extinct at the end of Paleozoic time.

unconformity—a substantial break or gap in the geologic record, caused by an interruption of deposition, by uplift and erosion, or by igneous activity.

uplift—a high area produced by movements that push rocks upward.

vein—a thin, sheetlike intrusion into a crevice, often with associated mineral deposits. Veins may also be composed of minerals deposited by groundwater.

vent—any opening through which volcanic material is ejected.

volcanic ash—fine airborne material ejected by a volcano.

volcanic dome—see **lava dome**.

volcanic neck—solidified lava that cooled within a volcanic conduit, generally more resistant than surrounding rock.

volcanic rock—rock formed from magma exploded from or flowing out on the Earth's surface.

wash—a Southwestern term for a normally dry stream bed.

weathering—changes in rock due to exposure to the atmosphere.

window—a hole in rock, well above surrounding surfaces, usually smaller than an arch or natural bridge.

Index

Page numbers in **boldface** indicate major discussions. Page numbers preceded by a "C" refer to the color section.

Other books from The Mountaineers include:

FRAGILE MAJESTY: The Battle for North America's Last Great Forest, by Keith Ervin
In-depth examination of all sides of the controversy surrounding preservation of old-growth forests in the Pacific Northwest. A story about people as well as trees, and the uncertain future they share. Cloth, $22.95; paper, $14.95

TIMBERLINE: Mountain and Arctic Forest Frontiers, by Stephen Arno and Ramona Hammerly
Explains the timberline phenomenon in North America. For amateur naturalists and serious students. Paper, $14.95

FIELD GUIDE TO THE CASCADES AND OLYMPICS, by Stephen Whitney
Describes, illustrates over 600 species of plants and animals found in the mountains from Northern California through Southwestern British Columbia. 36 color, 53 black-and-white plates. Paper, $16.95

HIKING THE SOUTHWEST'S CANYON COUNTRY, by Sandra Hinchman
Six two- to three-week itineraries of dayhikes, backpacks and scenic drives in Colorado, Utah, New Mexico, and Arizona. Includes distance, duration, topographical maps, difficulty, and more. Paper, $12.95

MOUNTAIN FLOWERS, photos by Ira Spring, text by Harvey Manning
Pocket-sized field identification guide to 84 of the most common wildflowers of the Cascades and Olympics. 84 full-color photos. Paper, $6.95

PREGNANT BEARS AND CRAWDAD EYES: Excursions & Encounters in Animal Worlds, by Paul Schullery
Funny, informative, sometimes bizarre essays revealing the secret lives of animals. Gives an entertaining view of how they are affected and perceived by humans. Cloth, $18.95; paper, $12.95

A WILDERNESS ORIGINAL: The Life of Bob Marshall, by James Glover
Richly textured biography of a giant (and perhaps the only humorist) of the conservation movement, and founder of The Wilderness Society (1936). Marshall was also an author of note, and the namesake of the rugged Bob Marshall Wilderness in Montana. Cloth, $17.95

ANIMAL TRACKS OF THE SOUTHWEST, by Chris Stall
Tracks and information on animals and birds of the region. Books covering other areas are also available. Paper, $5.95

Ask for these at your book or outdoor store, or order from us toll-free with VISA or MasterCard by phoning 1-800-553-4453. Mail order by sending check or money order (add $2.00 per order for shipping and handling) to:
The Mountaineers
1011 S.W. Klickitat Way, Suite 107, Seattle, WA 98134

Ask for free catalog of over 200 outdoor titles

Own the entire "Pages of Stone" series:

1: ROCKY MOUNTAINS & WESTERN GREAT PLAINS

National Parks: Badlands, Glacier, Grand Teton, Rocky Mountain, Theodore Roosevelt, Wind Cave, Yellowstone
National Monuments: Agate Fossil Beds, Craters of the Moon, Devils Tower, Dinosaur, Florissant Fossil Beds, Fossil Butte, Great Sand Dunes, Jewel Cave, Timpanogos Cave

2: SIERRA NEVADA, CASCADES, & PACIFIC COAST

National Parks: Channel Islands, Crater Lake, Lassen, Mount Rainier, North Cascades, Olympic, Redwood, Sequoia, Kings Canyon, Yosemite
National Monuments: Cabrillo, Devils Postpile, John Day Fossil Beds, Lava Beds, Mount St. Helens, Oregon Caves, Pinnacles

3: THE DESERT SOUTHWEST

National Parks: Big Bend, Carlsbad Caverns, Guadalupe Mountains
National Monuments: Bandelier, Capulin Mountain, Chiricahua, Death Valley, Gila Cliff Dwellings, Joshua Tree, Lehman Caves, Montezuma Castle and Tuzigoot, Organ Pipe Cactus, Saguaro, Salinas, Tonto, White Sands

4: GRAND CANYON & THE PLATEAU COUNTRY

National Parks: Arches, Bryce Canyon, Canyonlands, Capitol Reef, Grand Canyon, Mesa Verde, Petrified Forest, Zion
National Monuments: Black Canyon of the Gunnison, Cedar Breaks, Canyon DeChelly, Colorado, El Morro, Natural Bridges, Navajo, Rainbow Bridge, Sunset Crater, Walnut Canyon, Wupatki, and Chaco Culture National Historical Park

THE MOUNTAINEERS

1011 S.W. Klickitat Way
Seattle, WA 98134

Order toll-free (with VISA or MasterCard only):
1-800-553-4453

Call or write for illustrated catalog of more than 200 outdoor, nature titles published by The Mountaineers

Send me:

_____ Pages of Stone 1: Rocky Mtns/Great Plains @ $14.95 ea

_____ Pages of Stone 2: Sierra Nevada/Cascades/Pacific @ $16.95 ea

_____ Pages of Stone 3: Desert Southwest @ $16.95 ea

_____ Pages of Stone 4: Grand Canyon/Plateau @ $16.95 ea

☐ Free catalog of over 200 Mountaineers outdoor titles

I enclose $_____ (Add $2.00 shipping charge. Wash. Res. add 8.2% tax)

Name _____

Street _____

City _____ State/Zip _____